1007892548

Science Without Numbers, Second Edition

Science Without Numbers

A Defense of Nominalism

Second Edition

Hartry Field

OXFORD
UNIVERSITY PRESS

OXFORD
UNIVERSITY PRESS

Great Clarendon Street, Oxford, OX2 6DP,
United Kingdom

Oxford University Press is a department of the University of Oxford.
It furthers the University's objective of excellence in research, scholarship,
and education by publishing worldwide. Oxford is a registered trade mark of
Oxford University Press in the UK and in certain other countries

© This edition Hartry Field 2016

The moral rights of the author have been asserted

First Edition published in 2016

Impression: 2

All rights reserved. No part of this publication may be reproduced, stored in
a retrieval system, or transmitted, in any form or by any means, without the
prior permission in writing of Oxford University Press, or as expressly permitted
by law, by licence or under terms agreed with the appropriate reprographics
rights organization. Enquiries concerning reproduction outside the scope of the
above should be sent to the Rights Department, Oxford University Press, at the
address above

You must not circulate this work in any other form
and you must impose this same condition on any acquirer

Published in the United States of America by Oxford University Press
198 Madison Avenue, New York, NY 10016, United States of America

British Library Cataloguing in Publication Data
Data available

Library of Congress Control Number: 2016945126

ISBN 978-0-19-877791-5 (hbk.)
     978-0-19-877792-2 (pbk.)

Printed in Great Britain by
Ashford Colour Press Ltd.

Contents

Contents New to this Edition

Preface to Second Edition   P-1

0.1. Arithmetic and Cardinality Quantifiers   P-8
0.2. Mereology and Logic   P-12
0.3. Representation Theorems   P-14
0.4. Conservativeness   P-16
0.5. Indispensability   P-30
0.6. Other Forms of Anti-Platonism   P-38
0.7. Miscellaneous Technicalia   P-39

Bibliography for Second Preface   P-51
Note   P-54
Letter from W. V. Quine   P-55

Contents of First Edition

Preface to First Edition   i

Preliminary Remarks   1

1. Why the Utility of Mathematical Entities is Unlike the Utility of Theoretical Entities   7
   Appendix: On Conservativeness   17

2. First Illustration of Why Mathematical Entities are Useful: Arithmetic   22

3. Second Illustration of Why Mathematical Entities are Useful: Geometry and Distance   26

4. Nominalism and the Structure of Physical Space   31

5. My Strategy for Nominalizing Physics, and its Advantages   42

6. A Nominalistic Treatment of Newtonian Space-Time   47

7. A Nominalistic Treatment of Quantities, and a Preview of a Nominalistic Treatment of the Laws Involving Them   55

8. Newtonian Gravitational Theory Nominalized — 61
   A. Continuity — 61
   B. Products and Ratios — 64
   C. Signed Products and Ratios — 68
   D. Derivatives — 70
   E. Second (and Higher) Derivatives — 73
   F. Laplaceans — 76
   G. Poisson's Equation — 78
   H. Inner Products — 81
   I. Gradients — 84
   J. Differentiation of Vector Fields — 85
   K. The Law of Motion — 88
   L. General Remarks — 89

9. Logic and Ontology — 92

Bibliography for Original Text — 107
Index (to entire volume) — 109

Preface to Second Edition

When I began writing *Science Without Numbers* in the winter of 1978/79, I did not intend to write a book, but a long article; but it grew until it reached a point where publication as a normal journal article did not seem feasible. And the final product didn't really seem like a book (which is why I called it a monograph, a term I never used before or since): I gave only a cursory motivation for a certain project, that of presenting physical theories in a certain ("nominalistic") format, and spent the rest of the time trying to overcome skepticism about the feasibility of the project by giving a detailed sketch of how it might be accomplished for a particular non-trivial physical theory.[F1] A "real book" would have required a more detailed philosophical discussion of the motivations for the project, but I wasn't ready to give that at the time of publication of the monograph. I did attempt more philosophical justification over the ensuing decade or so, in a number of articles, and in the Introduction to a volume (Field 1989/1991) that contained some of these articles. (I also devoted some attention to an obvious lacuna of the book, the issue of whether the nominalistic position of the book could accommodate metalogic.) By some time in the early 1990s, though, my interests within the philosophy of mathematics had shifted a bit,[F2] in part because of increasing doubts about the Quinean framework that *SWN* presupposed. In particular, I became increasingly doubtful of the following two suppositions:

- that the question of what exists has a univocal and non-conventional content;
- that the right way to answer this question is to look at the existential quantifications of our most fundamental theories; it is "doublethink" to employ a fundamental theory and not literally believe its posits if there is no serious prospect of showing how those posits could be eliminated.

[F1] "Defense" in the sub-title was intended literally: I was defending it against an attack, not going on the offense.
[F2] See for instance my 1994, 1998a, and 1998b.

I still regard myself as *anti-platonist* in a broad sense, and still regard the work in *SWN* as relevant to supporting the credentials of anti-platonism; but I no longer want to rest my anti-platonism on the claim that the program of *SWN* can be completely carried out.

'Platonism' can mean a number of things. In the book I took it to be primarily a thesis about *what exists*: in particular, I took the existence of mathematical entities to suffice for "platonism". "Nominalism" is the denial of platonism in this sense. But another interpretation of 'platonism' is that what distinguishes a good mathematical theory from a bad one is how accurately it describes mathematical reality. Slightly more precisely, the view (said to be "typical of extreme platonism" in Chapter 1 of the book) is that higher mathematics is objective in the way that the sciences are: not only is it objective what's a good proof, it's also objective what's a correct axiom, so that e.g. there's an objectively correct answer to the size of the continuum even though this is unsettled by current axioms. But platonism in this sense can go with anti-platonism in the "ontological" sense: witness the Putnam-Hellman idea (Putnam 1967; Hellman 1989) that mathematical questions are thoroughly objective but to be understood modally. Conversely, ontological platonism needn't require platonism in the "objectivity" sense: witness Mark Balaguer's "plenitudinous platonism" (Balaguer 1998), or the view that Putnam seems to advocate in the early section of Putnam 1980, or views on which the existence of mathematical entities and the laws they obey is a matter of convention.

There is considerable plausibility in the idea that the *arithmetic of natural numbers* is objective in the sense I've described, a fact that I take to be explainable by the close connection between it and the logic of cardinality-quantifiers. I think this is consonant with the viewpoint of *Science Without Numbers*, though the book ought to have emphasized it more. The book did, as noted, come out against this sort of objectivity for other parts of mathematics that don't seem so intimately related to logic; and in this I believe it was correct.

I now regard this objectivity issue as more important than the existence issue;[F3] indeed, I'm not entirely sure that the question of what exists has a univocal and non-conventional content, though I've never been entirely

---

[F3] " 'Ontology', I spoke the word, as if a wedding vow. Ah, but I was so much older then, I'm younger than that now." (Bob Dylan, approximately.)

satisfied by attempts like Carnap's (1950) to make sense of the view that it doesn't. But whatever one thinks of the existence issue, I think that the program of *SWN* bears on the objectivity issue (as it arises for parts of mathematics that go beyond the arithmetic of natural numbers), and tends to support the anti-platonist position on it. Moreover, if one does take the existence issue seriously, one can take *SWN* as bearing on it without buying into the Quinean view presupposed in the book, according to which existence questions are to be settled by reading the answers directly from our most fundamental theories. More on these matters later.

There were, actually, two main motivations for the project of *SWN*.

One concerned the platonism issues, most explicitly, the ontological one: the project was to rid us of the need to literally believe in mathematical entities (not just numbers). Normal ("platonistic") formulations of physical theories contain reference to all sorts of mathematical entities; whereas the "nominalistic" reformulations of theories that I proposed contained no reference to mathematical entities of any kind. The idea was that until you present a physical theory nominalistically, it looks as if literal belief in the theory requires literal belief in mathematical entities; but the possibility of nominalistic formulations shows this not to be the case. The view was that mathematical theories are essentially useful calculation devices. In the context of a given nominalistic theory, some consistent mathematical theories may be more useful for calculation than others, but this doesn't make them better in any context-independent or non-utilitarian sense. (The arithmetic of natural numbers is a partial exception since its most characteristic use is for dealing with cardinality quantification, a role that it has independently of any particular empirical theory.)

SWN is often described as advocating an *error theory* about mathematics, but I think that this description is highly misleading. The book does assert that mathematical theories aren't literally true, if taken at face value; but to say that this is an error theory suggests that most ordinary people, or mathematicians, or physicists who use mathematics in their theories, *falsely believe* the theories true in this sense. I am skeptical that ordinary people or mathematicians or physicists typically have any stable attitude toward that philosophical question. The only error I saw was on the part of platonist philosophers (e.g. my most explicit target, Quine, who often said that mathematical objects are real in just the way that physical objects are—see for instance Quine 1948).

John Burgess (1983; also Burgess and Rosen 1997) likes to distinguish between "revolutionary" and "hermeneutic" nominalists, the former trying to correct what they see as widespread error and the latter purporting to elucidate what ordinary people and mathematicians and physicists meant all along. In my view, this is a false dichotomy. I certainly didn't think that the account I was providing was "hermeneutic", but it wasn't "revolutionary" either: I took what I was doing, rather, as providing an account that explains why ordinary mathematical practice is perfectly fine, and doesn't require a platonist ontology. (The claim that my account accepts all of "ordinary mathematical practice" would have to be qualified if one construed ordinary practice as taking seriously questions like "What is the real cardinality of the continuum?". But of course the invention of new axioms and the investigation of their implications for the cardinality of the continuum is a part of mathematical practice that comes out perfectly fine on my account; it's only the question of which such competing axioms are "really true" that my account rejects.)

The other motivation for SWN was based on a feeling that though formulating an empirical theory using a high-powered mathematical apparatus can in many ways be illuminating (especially when it comes to comparing that theory with others), it can sometimes make it hard to see what is really going on in the theory. Formulating physical theories without the high-powered mathematical apparatus, and in what I called an "intrinsic" manner, was intended to illuminate those theories; and in combination with representation theorems, to give an account of the application of mathematics that would be appealing even to the platonist. The demands for a nominalistic formulation and for an "intrinsic" one aren't the same: a formulation of the theory of gravitation that made reference to a numerical inscription of 6.67×10^{-11} to represent the gravitational constant in $m^3/kg^{-1}/s^{-2}$ might count as nominalistic, but would seem far less "intrinsic" than some formulations that aren't fully nominalistic. This second motivation for my program started out as about as important to me as the first, but I never managed to make it or the idea of intrinsic explanation very precise, and the philosophical justification that I embarked on in the ensuing decade was devoted almost exclusively to the first motivation. (A partial exception is the discussion of the distinction between "heavy duty platonism", "moderate platonism", and "very moderate platonism" in Field 1985b.)

Besides containing only cursory philosophical discussion, *SWN* was incomplete as even a sketch of how to avoid literally believing in mathematical entities. While it suggested a method of formulating (certain kinds of) fundamental physical theories in a way free of reference to mathematical entities, it said nothing about the use of mathematical entities in other areas, in particular in using metalogic to facilitate logical reasoning. What the nominalist should say about metalogic is a matter I turned to in the next decade or so, in particular in Field 1984, 1991, and 1992. And that work is indirectly relevant to the case of fundamental physical theories. For as David Malament (1982: 528–9) noted in an excellent critical review of *SWN*, much discussion of such theories concerns the existence of models with certain features, or (as in the case of determinism) whether all models with certain features have certain other features, and so on: in short, it concerns claims about what is consistent with the theory, or what follows from it given certain assumptions. I certainly think such questions are sensible; my view of them is that the model-theoretic formulations are "abstract counterparts" of formulations in terms of logical possibility, which I argued to be intelligible independent of mathematical entities (and to be a clearer and more austere notion than "metaphysical possibility"). *SWN* also said nothing about the use of mathematics in sciences other than fundamental physics, e.g. economics or psychology. My view was that these theories are heavily idealized anyway, so not candidates for literal belief, so that if mathematical entities were indispensable in them it would be of no ontological significance. Still, I thought that even for idealized theories there is strong motivation for finding an "intrinsic" formulation over an "extrinsic" one (e.g. in Bayesian psychology, using relations of comparative credence rather than numerical credence functions); while I did not in any way contribute to that program, I had an interest in it.

The book dealt only with Newtonian gravitation theory, but it was perfectly clear that the basic ideas extend to other field theories in flat space-time, such as classical electrodynamics (viewed special-relativistically). A number of people have suggested that there would be a problem extending it to general relativity. This surprised me: the methods that I used to handle quantities were clearly modeled after the treatment of co- and contra-variant tensors in differential geometry, and in the footnote at the end of Chapter 8, Section E I outlined a way in which one could apparently carry them over to curved space-times with an affine connection (i.e. with a notion of geodesic), or at least with a metric. As I said there, there are details that would need to be worked out (I give a *slightly* more complete sketch at the end of this Preface), and maybe there is a problem, but to my knowledge, none of the people who have expressed skepticism have given any hints as to where such problems might lie.

There do seem to be problems extending to certain other kinds of physical theory. In his aforementioned review, David Malament raised the problem of theories formulated in terms of configuration space or phase space: Lagrangian and Hamiltonian formulations of classical mechanics, Gibbs-style statistical mechanics, and quantum theory.[F4] My initial reaction to the examples of Lagrangian and Hamiltonian formulations was that these formulations are mathematically convenient and have heuristic value but can be thought of as instrumental. Perhaps one could say something like this of statistical mechanics too, without giving up the idea that, for truly basic theories, everything needed in the formulation of the theory must genuinely represent; though this doesn't strike me as entirely comfortable. But in the case of (even non-relativistic) quantum theory the problem seems very hard to escape: while there is presumably little difficulty in describing the wave function "intrinsically" in terms of predicates of comparative amplitude and comparative phase-difference,[F5] these would be predicates *on configuration space* (which is naturally viewed as the space of *possible configurations of particles* or *possible decorations of space*), or rather, on a space that adds to configuration space an extra dimension for time. There are ad hoc tricks that might allow recasting this in terms of predicates on ordinary space-time, but I don't know any that clearly work, and they would seem bound to make the description far more complicated.[F6]

[F4] A fuller discussion of the problem, with special reference to classical statistical mechanics, is in Meyer 2009.

[F5] Cian Dorr has pointed out to me that phase differences in quantum mechanics aren't invariant under Galilean transformations; as a result, the phase-difference comparison predicates need two extra places, for non-simultaneous points representing a state of motion.

[F6] Some of the discussion in the literature of the problem of carrying out the program of nominalization for quantum mechanics suggests that the problem is even worse, on the grounds that we would need to have a nominalistic analog of the algebra of Hermitian operators on Hilbert space. This is, at the very least, contentious: for instance, it doesn't arise on the view that "the operator observables of quantum mechanics are [merely] bookkeeping devices for effective wave function statistics" (Dürr and Teufel 2009: 228), a view which I find compelling independent of issues about nominalism. But even if the problem isn't worse, it is bad enough!

Independent of a commitment to phase space or configuration space, and indeed independent of issues of nominalism in general, there are other features of both classical statistical mechanics and quantum mechanics that suggest a partially instrumentalist treatment: I have in mind especially the notion of chance, whose role as simultaneously describing reality and directing our degrees of belief raises many philosophical perplexities.

These are among the pressures that eventually led me away from the strict Quinean standard that one's ontological commitments are to be read off one's views about the best ultimate theory, and toward the more relaxed ontological attitude to be suggested here.

I should mention that a number of people have made technical contributions to the positive program, most notably Frank Arntzenius and Cian Dorr in chapter 8 of Arntzenius 2012. (Among other things, they sketch an approach to general relativity that doesn't depend on the differential manifold having an affine connection, and so is presumably generalizable to other space-time theories in a way that my approach isn't; they also sketch an approach to fiber-bundle theories. In addition, they provide a good discussion of the philosophical significance of the program.) Brent Mundy has also contributed (1987 and elsewhere); his program allows quantification over physical properties, which I avoided, but I did say in Chapter 7 that quantification over physical properties would be "at least arguably nominalistic". John Burgess's 1984 also deserves mention, though the philosophical concerns behind it are rather different from mine: not only was he thoroughly opposed to nominalism, he also didn't share the (admittedly somewhat vague) desire to formulate laws "intrinsically". This is illustrated, for instance, by his heavy use of coding devices, and by his willingness to "quantify over arbitrary choices" of coordinate systems rather than avoiding the use of coordinate systems in the first place.[F7] (His claim in Burgess 1991 that space-time is every bit as dispensable as numbers depends on this.)

[F7] Like me, Burgess uses standard work in geometry and measurement theory to present the geometry and the "spaces" of physical quantities, and this is all as "intrinsic" in his case as in mine. He uses the coding devices to define the mathematics within the physics. My reservation about his approach is that when it comes to formulating the laws, he simply uses the standard mathematized formulations, understanding the mathematics as defined in this way; the formulations of the laws thus inherit the arbitrary coding built into his geometric definition of the mathematics. Similarly, the mathematized formulations depend on arbitrary choices, even if it is then shown that the choices don't matter.

SWN generated quite a lot of critical discussion. In part to avoid obviating any of this (but mainly because I'm far too lazy), I'm republishing the book without any substantive changes: I've merely corrected typographical errors and minor slips, and made a few other changes to improve clarity. For instance, I've broken up some overlong sentences and paragraphs, and improved the formatting for added readability; and I've switched from endnotes to footnotes, and put the references into a separate bibliography. Also I've done a bit of rewriting in a couple of proofs in Chapter 8 to make them easier to follow; and I've added two clarificatory footnotes, which I've marked as new to this edition. There are two slightly less minor slips in Chapter 1 concerning the definition of conservativeness (one in a footnote and the other in the Appendix) that I've treated specially: I've corrected them, but also noted the original wording. (I don't think anyone could have seriously been misled as to what was intended in either case, but I do know of one published discussion that rests on the original wording of the footnote, and it's possible that this is also so for the claim in the Appendix.) Similarly in Chapter 9, in formulating the first order nominalistic theory I omitted a needed axiom; I have added it, with a note on the addition, and done some rewording to accommodate its inclusion.

I have also decided against explicitly commenting in this preface on the critical discussions of the book, except in passing: it would be hard for me to do a good job after so many years away from these issues, and I fear that if I got in I could never get out.

But I will make a few general remarks about ways in which my thinking has evolved since I wrote the book. It should go without saying that my thoughts have doubtless been influenced by the literature and by comments I have heard over the years. I'm sure that were I to try to cite those who have influenced me I would come up with only a tiny proportion of those who belong on the list, so I will not try.

0.1. Arithmetic and Cardinality Quantifiers

One of the parts of the book I'm least satisfied with is the second chapter, on the arithmetic of the natural numbers. The view allowed us to regard sentences of form

There are exactly 87 *F*s

as literally true, since they could be paraphrased in terms of first-order logic with identity, quantifying over nothing but Fs (and in particular, not quantifying over numbers). But it did not allow us to regard sentences of form

There is a prime number of Fs

as literally true, because it could not be so paraphrased.

I now think the right response to this is that 'There is a prime number of' is a perfectly respectable quantifier in its own right. In the final chapter I contemplated adding some other quantifiers as primitive, e.g. the binary quantifier

There are fewer Fs than Gs.

But I now think it clear that a satisfactory theory must systematically add the means for defining a vast array of such quantifiers. Like Hodes 1984, I'm inclined to the following combination of views:

(1) Logic shouldn't postulate numbers.
(2) Logic should contain a rich theory of cardinality quantifiers—much richer than available in first-order logic.
(3) The primary point of the arithmetic of natural numbers is to encode these cardinality quantifiers, or at least some of them, and to encode the logical relations among them.

Arithmetic gives us a simple means for formulating the logic of such quantifiers; but at least when taken at face value, it does so by means of a fiction, or at least in a way that goes beyond the bounds of logic. The underlying idea is that because of the obvious ties between natural numbers and numerical quantifiers, the arithmetic of natural numbers is much closer to logic than other parts of mathematics are.

In carrying out this idea, it would be nice to formulate the logic of such cardinality quantifiers directly, without going beyond the bounds of logic. Exactly how this ought to go, I'm not sure. (I don't much like Hodes' way of doing it, which involves impredicative higher-order logic; I don't think the impredicativity sits well with his insistence that the "concepts" that the second order quantifiers range over are in some non-mathematical sense "predicative entities".[F8]) A first question is how vast an array of

---

[F8] If properties are "predicative entities" in Hodes' loose sense of being somehow like predicates, then the relation of *instantiating* or *falling under* should be "predicative" in a

quantifiers one wants to come out of it. I'd be inclined to want, at the very least, for each $k \geq 1$ and each primitive recursive k-place predicate Ψ, a k-ary quantifier Q_Ψ, where $(Q_\Psi x)[F_1(x) \ldots F_k(x)]$ means intuitively that $\Psi(\eta x F_1(x), \ldots, \eta x F_k(x))$ (where $\eta x F_i(x)$ is the number of x such that $F_i(x)$).[F9] And one might well think that the scope should include more than the primitive recursive. But whatever the decision on this, a second question is how best to formulate the theory so that it develops these quantifiers systematically from a small definitional base, without looking like it's relying on an ontology of numbers (or of anything else). This program strikes me as well worth pursuing.[F10]

One option, which seems more appealing to me now than it would have when I wrote *SWN*, is to partially mimic Hodes, but with some sort of *predicative* higher-order logic. (That's in a somewhat loose sense of 'predicative' that includes non-stratified approaches, and in particular includes the sort of nonclassical non-stratified approach to properties mentioned in note F8. These latter approaches don't actually restrict property-comprehension, but they are broadly predicative in that they

---

corresponding sense of being *like satisfaction*. And in that case, whatever precautions you think are needed to avoid semantic paradoxes are needed for properties too. And on almost every view of the semantic paradoxes, this gives rise to something like predicativity in the mathematical sense. (The most familiar classical view of the semantic paradoxes is Tarski's, according to which you have to stratify linguistic predicates; this corresponds almost exactly to the original technical notion of predicativity. Variant solutions tend to avoid the stratification, but something of the flavor of predicativity remains: e.g. on the property analog of the account of satisfaction in Kripke 1975 or its extension in Field 2008, while there are no predicativity restrictions on comprehension, still the law of excluded middle may fail for certain properties that are defined only impredicatively.)

[F9] As stated, the specification doesn't handle the case where for one or more of the i there are infinitely many F_i. What we really need to do is first extend the domain of application of Ψ from the natural numbers to the natural numbers plus a number "infinity" (which does not differentiate among infinite cardinalities); the intuitive meaning above is for the extended Ψ. (The notion of primitive recursiveness could be extended to the natural numbers plus infinity by means of the bijection taking infinity to 0 and each natural number to its successor. This would have the consequence that the finiteness quantifier counts as primitive recursive, which raises a question about how to understand the extended logic; I'll say a bit about this starting in the paragraph after next.)

[F10] I believe that one commentator on *SWN*, I can't remember who, suggested that I interpret such cardinality quantifiers in terms of numeral-shaped regions of space-time. I imagine that he or she was joking, but in case not: it is totally alien to the methodology of *SWN* to invoke space-time regions except in the context of theories that have space-time as their subject matter; and even there, it is totally alien to the methodology to use constructs that are "extrinsic" in a sense hard to precisely define but of which the proposal here would be a gross exemplar.

don't generally guarantee classical logic for impredicative properties.) Such a broadly predicativist approach wouldn't yield the logicist interpretation of arithmetic that Hodes wants (it wouldn't yield a translation of arithmetic language in which you can prove mathematical induction), but it would suffice for defining a vast array of numerical quantifiers and proving many properties of them.

But it's also worth exploring the possibility of developing a theory that gives you a large array of cardinality quantifiers without invoking even predicative higher-order quantification, by using some sort of inductive procedure for generating new cardinality quantifiers from some basic ones. (This would probably involve the use of schematic variables in the quantifier subscripts.) There are a number of *prima facie* possible ways to proceed here; I don't know whether any of them would ultimately yield a satisfactory logic.

The final chapter actually did make a suggestion in the direction of extending logic to allow for more cardinality quantification: it suggested the addition of a binary quantifer 'fewer than' (interpreted in such a way as to not make distinctions among infinite cardinalities). To this I expressed a somewhat ambivalent attitude. On the one hand I took it to be far more attractive than the introduction of second-order devices (including the "complete logic of Goodmanian sums", i.e. impredicative second-order mereology, on which more presently). On the other hand, after pointing out that a 'fewer than' quantifier enables us to define a finiteness quantifier Q_{fin} (or $\exists_{fin}$ as I somewhat inappropriately called it), I pointed out that if one defines logical consequence in the usual model-theoretic way this will lead to a consequence relation that is neither compact nor recursively enumerable, which may seem to go against the idea of logic. (The same would be true if one invoked a finiteness predicate of regions and took it to be, like '=', a logical notion.)

The latter consideration led me to tentatively propose dispensing with the finiteness quantifier, in the applications I put it to in Newtonian gravitation theory,[F11] by a predicate of regions, taken as non-logical. But there are better courses (that don't involve avoiding the use of the notion in the

---

[F11] As we'll see in 0.7(D), I could have avoided the use of the notion of finiteness there by an independently motivated expansion of the primitives. But the burden of this section is that even aside from the needs of the Newtonian theory, a treatment of cardinality quantifiers is highly desirable.

theory—see note F11). One is to keep the quantifier and come up with some kind of deductive system for it; exactly what that deductive system should involve depends on what general framework for defining cardinality quantifiers is best. But in any case, such a deductive system would presumably involve some sort of induction schema, and an adequate treatment of Q_{fin} as logical requires the extensibility of the schemas as the language expands (see subsection 0.4.3.). A variant procedure would be to use a predicate of regions (say a finiteness predicate) instead of the special quantifier, and axiomatize it with an induction schema, but treat the predicate as logical by understanding the schema to be extensible as the language expands. Either way, we could regard conclusions obtained by such a deductive system as valid, without commitment one way or the other to the question of whether conclusions involving Q_{fin} or the finiteness predicate that can't be established in this manner but which are valid according to the most obvious model-theoretic account ought to be given the honorific 'valid'.

A smaller point about this chapter is that, contrary to what it suggests, you can't really make it all that much easier to assess nominalistic inferences like the aardvaark inference *simply* by using arithmetic to code the quantifiers: to make the inferences manageable, you do need that, but you need some metalogic as well. This is one of several points in the book where work I did later, on how a nominalist should understand metalogic, is relevant to giving the full picture. But this doesn't affect the *main* point I was trying to make, which is that the use of arithmetic to code the quantifiers is perfectly legitimate from the nominalist viewpoint, given the conservativeness of impure number theory (to be discussed presently).

0.2. Mereology and Logic

The other way (besides the finiteness quantifier) that the middle chapters of *SWN* go beyond first-order logic is in connection with mereology, and I was far less happy with this. Indeed, I recognized that mereology as I was understanding it simply isn't part of logic if we understand logic as topic-neutral, for mereology as I understood it deals with special entities, space-time regions.[F12] For that reason, it has no use except when space-time is

---

[F12] There is an alternative attitude toward mereology, according to which any entities can be "summed" in a way that needn't have a spatial interpretation. I suppose that if one

the subject matter. (There is some precedent for use of the term 'logic' in connection with special subject matters—witness "temporal logic"—but on the whole I think that the extended use is unfortunate.) But if mereology is simply part of the theory of space-time structure, it doesn't seem legitimate to invoke a special and powerful consequence relation in connection with it. While I'm sure that I was aware of this, I'm embarrassed to say that the book doesn't address it head on. The last half of Chapter 9 does however express a preference for avoiding the special consequence relation, and making do with first-order logic, or perhaps first-order logic supplemented with the logic of the cardinality quantifier Q_{fin}. But this was left somewhat programmatic.

The last chapter does not sufficiently emphasize the difference between going beyond first-order logic in connection with mereology and going beyond it in connection with cardinality quantification. I've already mentioned one point: cardinality quantification is topic-neutral, whereas mereology (as I construe it) isn't. But there's a related point, which is how we understand schemas. If a physical theory says something about regions definable in the language of that theory, there is no obvious reason why it needs to say the corresponding thing about regions not definable in that language but definable in broader languages; because of this, when one posits regions via a comprehension schema, there is no reason to think that this involves an implicit commitment that when the language expands, the same comprehension schema together with the same assumptions about regions will be unrestrictedly valid. With cardinality quantifiers, the situation seems quite different I think that our understanding of finiteness does involve a commitment that the rules here hold not just in the case of language as it is currently but to any expansion of the language. (Certainly other logical schemas behave like this: we don't normally think that the deMorgan laws are perfectly good logical laws for our current language but don't automatically hold for expansions of the language.)

---

accepted such a view, one could regard regions of space-time as atoms, and regard the "sum" of two perhaps overlapping regions r_1 and r_2 as an entity distinct from the sum of any other regions, and from the smallest region that includes all spatial parts of r_1 and r_2. That would yield far more expressive power, but it strikes me (and struck me then) as philosophically unpalatable: it would just be monadic second-order logic (of a presumably impredicative sort) by another name.

I'll say more about these issues after I've discussed why extensions of first-order logic might seem needed, which is in connection with representation theorems.

0.3. Representation Theorems

The reason I found it necessary (at least temporarily) to employ the "complete logic of mereology" was for the Hilbert representation theorem that I discussed in Chapters 3 and 4 and extended in the chapters that follow it. And these later extensions of the representation theorem in this form seem enough to satisfy a reasonable demand for "intrinsicness", even if the "complete logic" is thought of as depending on set theory. But of course I wanted a formulation that was genuinely nominalistic, and for that purpose any dependence on set theory was *verboten*.

However, as I mentioned in passing in Chapter 4, more general representation theorems are available, where the representation isn't necessarily by the real numbers but by some real-closed field. Tarski 1959 gives a representation theorem for a purely geometric theory in which there is quantification only over points; it allows representation by an arbitrary real-closed field. If one extends the underlying geometry to include regions, with a first-order axiomatization of them that doesn't rely on sets, then only rather special real-closed fields are representing fields. (That's because the comprehension schema for regions will guarantee the existence of regions that are specified via impredicative quantification over regions, and because Dedekind completeness (which can now be given in a single axiom instead of a schema) will then guarantee that if bounded these have closest bounds.) This more general sort of representation theorem can be used for physics more generally; I'm not sure that I was fully aware of this in the book (though perhaps there are hints in Chapter 9), but this was a major theme of a follow-up paper (Field 1985a). The extended representation theorem for physics given in that paper says that for any model M of the nominalistic physical theory N, there is a real-closed field F_M, depending on M, whose "real numbers" can represent distance and the various physical quantities. (The same representing field is used for both.) Again only rather special real-closed fields are representing fields, not only for the reason given above but also because the physical vocabulary of the theory can be used to specify bounded sub-regions of lines, and the Dedekind completeness axiom for regions then guarantees that these too have closest bounds. So F_M will be "close to the real numbers". (If the quantifier Q_{fin} is used in N, and the model M is assumed to treat it standardly, then F_M is Archimedean, so can be taken to be a subfield of the reals.)

I think I ought to have used representation theorems of this more general sort in the book. (Mundy 1987 also advocates their use.) It wouldn't have changed a whole lot in the position advocated, but it would have made the first-order option discussed in Chapter 9 more appealing. (Or the "almost first-order" option also mentioned there, of using Q_{fin} but nothing beyond that.) Of course, as with any version of the first-order (or "almost first-order") option, it would mean that the proposed nominalized physics doesn't capture quite the full content of standard platonistic physics. I doubt that the loss would be of much significance to physics: work on subsystems of second-order arithmetic (see Simpson 1999 for an excellent survey) suggest that far less than the full content of even Henkin second-order arithmetic plays a role in physics. (In the unlikely event that some aspect of standard platonistic physics not captured in the proposed nominalistic axiomatization proved important, we could look for a richer nominalistic theory that did capture it: it's not as if this part of mathematics is *in principle* inapplicable.) It would be nice to know just where the mathematics used in the platonistic theories that one gets from the representation theorems mentioned (the one without Q_{fin} and the one with it) fit with the hierarchy of subsystems of second-order arithmetic, but that is better addressed by others. (The answer might well be affected by the need, noted in 0.7(A), to include a nominalistic analog of integration theory in the nominalistic system.)

We could also avoid the use of representation theorems altogether. This is what is proposed in Burgess 1984 and Burgess and Rosen 1991: instead of representation theorems, we develop a piece of mathematics (a subsystem of second-order number theory) within a very simple conservative extension of the synthetic theory, one obtained by adding certain equivalence classes. (For instance, we identify real numbers with equivalence classes of ratios of line-lengths.) I'm not sure that this is ultimately very different from the approach of the previous paragraph, but it gives a different emphasis in three respects. First is their de-emphasis of intrinsicness in the formulation of laws, a matter I've already discussed. Second, their approach encourages the view that we're in a fairly literal sense defining the mathematics geometrically, and that once so defined we're free to use it in applications (say in the theory of credences); whereas part of my picture was that if ratios of line-lengths aren't "intrinsic to" credences we shouldn't use real numbers so defined for credences. Third, their approach doesn't emphasize that the mathematics that can't be so defined, such as higher set theory, can be legitimately employed in geometry and elsewhere, by the conservativeness property that I now turn to.[F13]

0.4. Conservativeness

One of the points on which the book was widely criticized, and to some degree justifiedly, was the discussion of "conservativeness" in the opening chapter. I stand by the basic idea, but there are some things I wish I had done differently.

0.4.1. The Basic Idea

The idea was to state a criterion of goodness for mathematical theories that doesn't involve truth. In some ways conservativeness was to be stronger than truth: it was intended to capture the idea that whether a mathematical theory is good is independent of what the physical world is like. It is often assumed that the way to capture this idea is to say that mathematics is *necessarily true*, but my criterion of conservativeness was to be "like necessary truth but without the truth". An informal characterization, close to what I gave, would be this:

> A mathematical theory S is *conservative* iff for any nominalistic assertion A, and any body N of such assertions, A isn't a consequence of N + S unless A is a consequence of N alone.[F14]

("N + S" simply meant the union of N and S, considered as sets of theorems. In subsection 0.4.3 I'll look at the charge that this gives too anemic a reading of what it is to "add" a mathematical theory to a nominalistic one.) The claim was that good mathematical theories are conservative in this sense.

[F13] As we'll see, its use *in empirical comprehension principles* goes beyond conservativeness.

[F14] Both this definition and the variant in the book are in platonistic terms: I left the task of "nominalizing" metalogic to later work.

An equivalent formulation invokes the notion of consistency according to which a theory is consistent iff not everything is a consequence of it (or if no contradiction is a consequence of it): then a conservative theory is one that is consistent with every consistent nominalistic theory, i.e.

A mathematical theory S is *conservative* iff for any consistent body N of nominalistic assertions, N + S is also consistent.

There is a boring issue to be addressed about the reading of 'nominalistic assertion'. In the book I counted claims like 'Everything has mass' and 'There are fewer than $10^{10^{10}}$ things' as nominalistic, on the grounds that they don't imply the existence of mathematical objects as normally conceived. But not only don't they imply the existence of mathematical entities, they pretty much rule them out: more precisely, the second of the two claims is inconsistent with the existence claims of standard mathematics, and the first is inconsistent with the usual conception of mathematical entities and hence with what would seem to be a harmless supplementation of standard mathematics. And because of this, conservativeness *as formulated here* (with N + S taken as simply the union of N and S) would come out trivially false for standard mathematical S, on that understanding of 'nominalistic assertion'. There are two ways to avoid the problem. One way keeps this formulation of conservativeness but takes 'nominalistic assertion' more narrowly: it requires a nominalistic assertion to be *neutral to* the existence of sets or numbers or other sorts of mathematical entities, i.e. they not only can't imply their existence but also cannot rule them out. The obvious way to achieve this is to suppose that, in a nominalistic assertion, all quantifiers must be restricted to non-mathematical objects; and I've used that understanding in various papers subsequent to the book. The second way to avoid the problem, the one adopted in the book, keeps the broader understanding of 'nominalistic assertion', but replaces the formulation of conservativeness with something more complicated that makes appropriate quantifier restrictions to achieve neutrality (and handles a complication about an existence assumption built into standard logic). There is nothing of any philosophical significance in the divergence between these approaches, they're simply two different ways of doing the same thing. (They give rise to a further minor difference in the technical formulation of conservativeness, to be addressed later.)

Of more importance is what is meant by 'mathematical theory'. Obviously I didn't mean 'theory that uses mathematics (or mathematical entities) in its formulation', for standard formulations of physical theory do that, and they have nominalistic consequences on their own and hence certainly aren't conservative. Rather, I meant 'theory that should be regarded as part of mathematics'. This includes *impure mathematical theories* as well as pure ones.

A *pure* mathematical theory doesn't speak at all about non-mathematical entities, and its only non-logical vocabulary is special to mathematics; such theories (for which conservativeness reduces to consistency) are of interest to applications only as parts of larger mathematical theories. The larger mathematical theories are *impure*, like impure set theory or impure number theory; and it is these for which conservativeness strengthens consistency.

- I take *impure set theory* to

    (1) Not only posit "pure" sets but conditionally posit sets of physical objects: the theory says that for any physical objects there may be, there are lots of sets that have them as members—including a set of all of them. (Also of course, sets *of sets* of physical objects, sets that include both pure sets and physical objects, and much more.)
    (2) Allow the vocabulary for physical objects to be used in specifying what sets there are. For instance, in impure set theory based on a language that includes 'star', we may speak of the set of all stars. (Without allowing words such as 'star' to appear in the comprehension principle, this is a set which wouldn't be definable even with parameters, if there are infinitely many stars.)

Despite these features, impure set theory is part of mathematics: a platonist would regard it as true by mathematical necessity. A somewhat similar impure theory (though simpler because it doesn't posit objects not in the corresponding pure theory) is impure number theory.

- *Impure number theory* includes an operator 'the number of' which can apply to formulas that include non-mathematical vocabulary, so that we can say such things as "For any star x, if there are exactly two planets of x then the number of planets of x is prime".

I didn't specifically talk about impure number theory in the book—there was no real need to, since it is a consequence of impure set theory and so the latter can serve all the purposes that the former can—but it is another example of the kind of impure theory I had in mind. Theories like these, even though impure, are the sort of things that a platonist would regard as true by mathematical necessity; and they are the kind of theory I held to be conservative. By contrast, a claim like 'Every object is located in Euclidean 3-space' isn't part of mathematics but of physics: no platonist would regard it as true of mathematical necessity, and the conservativeness condition quite properly rules it as not part of what we should regard as good mathematics.^{F15}

There seems to be some confusion in the literature between on the one hand my conservativeness claims (that good mathematics is conservative, and that standard impure number theory and impure set theory are good in this sense), which I expected would be uncontroversial once pointed out, and on the other hand certain claims related to the dispensability of mathematical entities, which I took as far less obvious. These less obvious claims concern the existence of suitable nominalistic theories. *Once one has an interesting nominalistic theory T_0*, the conservativeness of impure set theory tells us that the result of adding impure set theory to it adds no new nominalistic consequences. So if it can also be shown that the result of adding impure set theory to it has pretty much the same nominalistic consequences as a platonistic theory T, then conservativeness entails that T_0 and T have pretty much the same nominalistic consequences, which is what one needs for the dispensabilty of mathematical entities for T. But the conservativeness claim does nothing to show the existence of the interesting nominalistic theories.^{F16} My apologies if I'm belaboring the obvious, but even as astute a philosopher of mathematics as Michael Dummett seems to have been confused by the point. In Dummett 1994 he writes,

Field envisages the justification of his conservative extension thesis as being accomplished only piecemeal. For each mathematical theory, and each theory to

---

^{F15} A more "mathematical-looking" theory that is ruled out of good mathematics by the conservativeness criterion is the modification of impure set theory obtained by replacing the axiom of infinity by its negation (keeping the replacement schema and the existence of a set of all non-sets): if it includes the axiom of choice it rules out all theories in which the physical world is infinite, and even without choice it rules out most, e.g. those with a definable infinite linear order relation.

^{F16} Of course, each platonistic theory T conservatively extends the "theory" consisting of the Craigian transcription of the set of nominalistic consequences of T; but I assume that everyone agrees that such Craigian theories aren't interesting.

which it is to be applied, the demonstration is to be carried out specifically for those two theories; no presumption is created by the successful execution of the programme for one case that it will work in others. (p. 17)

This is incorrect: I gave reasons (indeed, platonistic proofs of a sort) for thinking that impure set theory is generally conservative, from which it follows that weaker theories are too. For instance, consider the theory obtained by taking the theory of real numbers and adding to it just enough impure set theory to speak of functions from physical objects to the real numbers, so that we can apply the real numbers to the physical objects. This is guaranteed to apply conservatively to any nominalistic theory whatsoever, as a result of the fact that impure set theory does and that this theory is a subtheory of that. Dummett does have a point here, but it isn't about conservativeness: rather, it is that finding an interesting nominalization of one physical theory by no means guarantees that one can find it for another. (Even so modified, his 'no presumption' claim is a bit strong: to the extent that the theories are similar in mathematical structure, I'd think that nominalization of one is grounds for expecting that we could nominalize the other. But it is right that there is no guarantee.)[F17]

I've been spelling out the intuitive idea of conservativeness; but there are some issues that need to be clarified. I will discuss the main issues in the next subsection. But first I should make more explicit one other technicality in the definition of 'nominalistic assertion', especially since the definition as given in footnote 8 of the original edition included a slip. If we want to use the simple formulation of conservativeness, we need that a nominalistic assertion (in a language with no singular terms other than variables, to make things simple) is one:

[F17] It's possible that rather than confusing conservativeness with nominalizability, Dummett was misunderstanding how conservativeness was supposed to work. Perhaps he thought that the impure mathematical theory containing real numbers that we add to certain nominalistic theories didn't consist in simply the theory of real numbers plus a theory of functions from objects to numbers *that follows from impure set theory*, but instead was a theory with substantial physical presuppositions that might be met for some physical theories but not for others. But I'm not sure how to fill out such an alternative interpretation in detail, and I think it should have been clear from the book that whatever substantial physical presuppositions are needed for an application of the real numbers need to be built into the nominalistic theory rather than be taken as part of the mathematics. So again, whatever objection there might be in the vicinity of Dummett's remarks, it is not to the conservativeness claim I was defending.

(A) in which all quantifiers are restricted to non-mathematical objects, and
(B) which employs no specifically mathematical vocabulary. (Logical vocabulary, including '=', is of course allowed; so is a special term 'mathematical', used to make the restriction to non-mathematical objects explicit.)

[(B) is important: without it, the claim (*) "there are non-mathematical x and y such that $x \in y$" would count as "nominalistic". But impure set theory says that anything that has a member is a set, and it was essential to build into impure set theory that sets are mathematical (in order that the quantifier restrictions to non-mathematical objects serve their purpose). So (*) is inconsistent with impure set theory, and that would violate conservativeness if (*) were counted nominalistic. This is for the simple formulation of conservativeness. For the one in the book, (A) can be dropped, since the restriction of variables is done by other means; but (B) is still needed, and indeed must be strengthened to preclude "mathematical" from appearing in nominalistic statements, or at least to include restrictions on the kind of occurrences it can have in them.[F18]]
Unfortunately in that footnote 8 I slipped in the formulation of (B), and said that to be nominalistic, a statement must not employ any non-logical vocabulary *that appears in our mathematical theories*. This made no sense in the context, where I'd stressed that all non-mathematical vocabulary is part of the impure mathematical theories of interest; and none of the subsequent discussion depended on it. I'm sure no one was misled, but I've corrected it in the current edition, with a statement about the change to the original text.

[F18] Thanks to Marko Malink for the observation that it must be strengthened in this way. Illustration: In Principle C of Chapter 1, take N to be the rather trivial theory $\forall x(x = x)$, and A to be 'Everything is non-mathematical'; A doesn't follow from N, but A^* is vacuous. (The last sentence of the derivation of Principle C from Principle C' in the first paragraph of note 15 of the text relied on the assumption that nominalistic claims don't include 'mathematical'.) The blanket refusal to let any sentence containing 'mathematical' count as nominalistic is slightly counterintuitive, in that no statement of form A^* where A contains quantifiers would count as nominalistic on this criterion. Perhaps it would be more natural to disallow the use of 'mathematical' in nominalistic claims except in the contexts 'for all non-mathematical $x \ldots$' and 'for some non-mathematical $x \ldots$'. Or we could avoid the whole issue by using the simpler sort of formulation of conservativeness given above.)

0.4.2. Consequence, Proof, and ω-Conservativeness

In the Preliminary Remarks and the opening chapter I tacitly assumed that we were working in first-order logic. I wrote these parts of the book early on, but as the book developed I seriously contemplated expanding the logic of the theory.

One expansion was to something akin to "full" monadic second-order logic, since this was used in the Hilbert representation theorem. I rejected ordinary second-order logic on the grounds that the second-order variables are required to range over sets or Fregean concepts or some such things. (This understanding of second-order variables has recently been challenged—see Rayo and Yablo 2001—though in my opinion the worries about impredicativity that I voiced earlier arise as well for the Rayo-Yablo view, and predicative second-order logic wouldn't suffice for the representation theorem.) In a desperate attempt to keep the Hilbert representation theorem nonetheless, I suggested a less general version of monadic second-order logic that I called "the complete logic of mereology" [or "the complete logic of Goodmanian sums", which was supposed to make it sound more nominalistically respectable]. The idea was to suppose, *as a matter of logic(!)*, that the regions of spacetime form a complete atomic Boolean algebra except for the absence of an empty region. Then quantifying over regions is a surrogate for quantifying over sets of atoms of the algebra, i.e. over sets of space-time points. This is less than one would get with monadic second-order reasoning generally: since regions were taken as basic entities, monadic second-order logic would have variables ranging over sets or concepts of arbitrary regions, not just of the atomic regions that are points. But I didn't need that added power for any representation theorems (or anything else), so I thought the "complete logic of mereology" would do.

But as I said in section 0.2, the idea of a "logic" of mereology seems misguided. I regret having considered it. I also still tend to resist the more genuinely second-order option, especially in the impredicative and indeed non-axiomatic form that would be required for Hilbert's version of the representation theorem.

The other expansion I contemplated, which played a far more limited role though which I was much happier with, was to include the cardinality quantifier 'fewer' or 'finitely many', conceived as logical.[F19] Given second-order logic, the cardinality quantifiers are definable (and given the "complete logic of mereological sums" they are definable *as applied to points of space-time*); nonetheless in Chapter 9 I considered the two expansions independently, so as to allow for the option of using the cardinality quantifiers without second-order logic or its mereological analogue. (This option was emphasized more emphatically in Field 1985a.) As I've noted, this would allow the use of representation theorems in which the representing field is guaranteed to be Archimedean.

As I said, the early parts of the book were written before I appreciated the pressures toward going beyond first-order logic. (When those pressures did become evident, instead of rewriting the early material as I should have, I simply made a few inadequate remarks at the end of Chapter 4 and in Chapter 9 about how what was said earlier ought to be adjusted.) In a more-than-first-order setting, there is room to ask what is meant by 'consequence' as it occurs in the definition of conservativeness.

If logic is taken to be thoroughly first order, there is no real issue about the extension of the consequence relation: we all agree that first-order consequence is syntactically characterizable. (At least, it is as long as we put issues of achieving a nominalistic metalogic aside, as I did in the book and I will continue to do here.) As long as we're happy with the purely first-order extended representation theorem, there isn't even a *prima facie* issue about what 'consequence' means in the definition of conservativeness.

But what if we consider the cardinality quantifier 'fewer' or 'finitely many', or second-order quantifiers, or mereology, as logical? In each case, the obvious semantic definition of consequence (in terms of standard models) gives a relation that is neither compact nor recursively enumerable; it would thus diverge from a syntactic definition in terms of proofs. Of course as a nominalist I wasn't going to *define* consequence either in terms of models or in terms of proofs, but there was still the question

---

[F19] I didn't consider the option of adding such quantifiers as new syntactic operators without considering them as logical: I assumed that if you didn't want them as logical you would dispense with them in favor of predicates. (Conversely, I didn't consider the issue of adding a predicate but treating it as logical, in the manner mentioned above in section 0.1 and discussed further below.) In retrospect the issue of operators vs. predicates seems inessential to the points at issue.

of whether on my understanding it lines up in extension with (a) the extension that it would have on a semantic construal were platonism true; (b) the extension it would have on a syntactic construal were platonism true; or (c) something else, presumably something in between. (Or maybe its extension is somewhat indeterminate.)

In note 27 of Chapter 4 I took a stand for its lining up with the semantic (and explained the apparent syntactic remarks in the opening chapter as due to a temporary tacit assumption of first-order logic, where the two coincide). But *at least in the case of full second-order logic or the "logic" of mereology*, this stand makes much of the rhetoric in the opening chapter quite misleading. The reason is that, as many people correctly complained later (e.g. John Burgess and Stewart Shapiro), semantic conservativeness simply amounts to truth in this second-order context. Indeed I now think that we have no clear grasp of second-order consequence (or its mereological analog) in the sense intended in the book, i.e. the semantic consequence relation of "full" impredicative second-order logic; I think the book takes this notion entirely too uncritically at a number of points.

On the other hand, the idea of a logic of cardinality quantifiers such as Q_{fin} seems much more appealing. For it too, the obvious semantic characterization diverges from any syntactic one, and so the issue of how 'consequence' is to be understood arises in this context as well. And it seems to me that if we're to regard this as a logical notion, we can't suppose that our grasp of the induction principle is adequately captured by the totality of instances of the induction schema in any fixed first-order language. The usual alternative to such a syntactic characterization is a semantic characterization in terms of standard models. Of course, a nominalist can't literally embrace a model-theoretic characterization, but a nominalist can't literally embrace a proof-theoretic characterization either; again, I want to put aside here the issue of how exactly a nominalist ought to deal with metalogic. My claim is just that the semantic characterization is a better platonist approximation to how we should think about the logic of Q_{fin} than is the syntactic characterization based on a fixed language.

If we do want to employ cardinality quantifiers in our nominalistic physics and think they have a logic that can't be syntactically characterized, to what extent would that undermine the general picture of the role of mathematics I tried to paint in Chapter 1?

I don't think it undermines the main idea: conservativeness in this sense (which to avoid any confusion might be called

ω-*conservativeness*)[F20] serves much of the function that necessary truth was supposed to serve, but doesn't imply truth. A mathematical theory that (ω-)implied that the physical universe was finite, or that it was infinite, or (ω-)implied any conclusion about the number of planets or the fate of the Paris Commune even if that answer were true, would be bad mathematics. But two mathematical theories that both met this condition might disagree about many purely mathematical claims such as the size of the continuum; short of positing an ambiguity they couldn't both be true, but on the ω-conservativeness criterion they could each be good mathematics. (Of course, there might be reasons why in some applications one was more useful; but which was the more useful might vary from one application to the next.)

The lack of synthetic characterizability may however undermine some of my rhetoric. For instance, in drawing the contrast between mathematics and physics I say that the conclusions we arrive at by applying mathematics to nominalistic premises "are not genuinely new, they are already derivable in a more long-winded fashion from the premises, without recourse to mathematical entities". This claim can be "saved" by replacing "they are derivable from" by "they are ω-consequences of", but I admit that this pulls at least some of their punch.[F21]

As mentioned, conservativeness is a slight generalization of consistency. Of course, if it's ω-conservativeness that's in question, this is consistency in (essentially) ω-logic, and thus not explainable in terms of any deductive system in a fixed language. I don't think this would seriously undermine the main point of Chapter 1. In the Appendix to that chapter I offer a (rather obvious) model-theoretic argument for why a platonist should believe that impure set theory (and hence any mathematics that

---

[F20] Incidentally, though I only came to realize it much later, the idea of ω-conservativeness is important in another context: in Kripke's theory of truth. The part of Kripke 1975 based on the Kleene evaluation schemes argues in effect that any theory adequate to basic syntax can be extended in an ω-conservative fashion to include a truth predicate in which (i) "True(<p>)" is everywhere intersubstitutable with "p" in non-intentional contexts, (ii) the usual composition rules hold, and (iii) 'True' is allowed in mathematical inductions. It's well known that the extension isn't deductively conservative (when the base theory is consistent and recursively enumerable): for the extended theory can prove the consistency of the base theory, whereas the base theory can't. So the focus on ω-conservativeness in the book wasn't really novel, though the use I put it to has a rather different flavor from what we have in Kripke's theory.

[F21] Even in the fully first-order case, "long-windedness" isn't really what's at issue; rather, the mathematical formulation is an *aid to seeing* what the consequences are.

can be formalized in it) is conservative. It goes through as before, except that we must start the construction from an ω-model of M of T; what the construction then produces is an ω-model of $ZFU_{V(T)} + T^*$. (Other arguments in the Appendix are more dependent on the logic being fully first order.)

0.4.3. Logic and the Extensibility of Schemas

It is built into my explanation of impure mathematical theories that any schemas that appear in such theories are to be interpreted as applying not just to instances containing only mathematical vocabulary and quantifying only over mathematical entities, but to instances containing physical vocabulary and/or quantifying over physical entities as well. (I don't rule out that in special circumstances there might be reason to introduce restricted schemas that impose limits on the vocabulary or quantifier-ranges of the instances, but one doesn't normally do so: the comprehension schema of impure set theory and the induction schema of impure number theory are supposed to apply to instances containing physical vocabulary, and it would cripple the application of mathematics to alter this.)

What about the reverse? Should schemas in empirical theories be deemed to extend to instances containing mathematical vocabulary? In the book I tacitly made a two-fold assumption about this.

The first half of the assumption was that, if the schemas appear in those theories as part of the logic, then they should be deemed to extend to all vocabulary including the mathematical. Logic, after all, is supposed to be topic-neutral.

The second half of the assumption was that any schemas that appear in physical theories that aren't based on logical principles (or on mathematical principles, in the case of platonistic theories) should *not* be extended to new vocabulary, without special justification for doing so. This seemed natural since such extension of the schemas adds empirical content. There could of course be special justification for extending the schemas. Most obviously, there might be empirical evidence for the added empirical content one gains from such an extension. Even without empirical evidence, considerations of simplicity or naturalness might perhaps favor the extension, though in the case of extension to mathematical vocabulary these considerations are likely to look different to the platonist than to the nominalist. (If one already has mathematical entities, then a theory that extends an empirical schema to instances that quantify over them or that contain special vocabulary for them might seem somewhat simpler or more natural than theories with a more restricted schema, though without empirical support for the excess content this wouldn't seem to me terribly weighty; but if one doesn't already have the mathematical entities, there is no such added simplicity, indeed there is added complexity in introducing the mathematical apparatus to extend the schemas.) But the view was that, without any such special justification, there is no reason to extend the schemas that appear in non-logical and non-mathematical theories.

To illustrate this two-fold assumption, and relate it to conservativeness:

(1) *If* one were to regard mereology as logic, or take N as a second-order theory, then we should view the schematic letters in the Dedekind-completeness schema or the comprehension schema for regions as in effect second-order variables, and as such, indefinitely extensible. In that case, if we add to N a set theory S that postulates lots of sets, then we should understand "N + S" as expanding the schema to include such sets; this would lead to things being provable in a deductive system for N + S that aren't provable in N without the expanded schema. This is no violation of conservativeness as I've explained it, because it depends on the assumption of second-order mereology or second-order logic more generally, where consequence exceeds provability. On this conception, unless S is a mathematically bad set theory, e.g. an inconsistent one, then N *already* implies whatever "N + S" does about the existence of regions; S merely serves to elucidate the meaning of the second-order quantifier and hence elucidate what were already consequences of N, in the non-axiomatizable sense of consequence in question.

(2) On the other hand, if N is taken as a merely first-order theory, I took it that there was no reason to expand the Dedekind-completeness schema when adding S to it: N is a theory about regions (i.e. parts of space-time) alone, saying nothing about sets, and there is no reason on this picture why sets are in any way relevant to the notion of region. For instance, even if the mathematics postulates non-measurable sets of points, there is no obvious reason why the physical theory must postulate regions corresponding to

them. (For instance, there is no obvious reason to regard a Banach-Tarski decomposition as physically meaningful.) So why should adding a claim about sets have any bearing on the understanding of the schema? (Of course if there were empirical support for the excess content that the platonistic theory with the extended schema has over the nominalistic one, the situation would be somewhat changed: the nominalist would want to try to attain this excess content without the platonism. But as I emphasized in the book, the excess content is quite *recherché* and currently has no empirical support.)

(3) Similarly, if N is taken as an "almost first-order" theory without second-order devices, but employing a quantifier Q_{fin} viewed as genuinely logical (or a finiteness predicate of regions, viewed as a logical predicate like '='), then the induction schema used to axiomatize this quantifier (or predicate) should be taken as indefinitely extensible, including to mathematical predicates when S is added to it. In that case, in analogy to 1, the addition of S to the theory will enable one to prove more from instances of the completeness schema that contain the quantifier (or predicate) than one can prove without S; but this is no violation of conservativeness, since it is merely elucidating what are already consequences of N. But since on this view there are no logical devices that go beyond first-order devices plus finiteness, there is no reason to expand the completeness schema to set-theoretic instances, absent empirical support for the expansion; in this respect the situation is as with 2.

In all three cases, "N + S" is just the union of N and S; it's just that N is conceived differently in the three cases.

Could any of this sensibly be denied? I doubt that anyone would want to deny that schemas for notions viewed as logical are indefinitely extensible, but I suppose that one might question the implicit assumption in 1 (and the first part of 3) that this extensibility goes all the way to instances that quantify over fictional objects or predicates appropriate to them. And if that assumption is wrong, then a nominalist should think that the indefinite extensibility doesn't extend to platonistic instances. If 1 were denied on this ground, then the "second-order nominalist" would presumably not be able to deduce anything that the first-order nominalist can't. The second order option would simply be irrelevant: we'd have deductive conservativeness even in the second-order case, but no representation theorems stronger than those in the first-order theory. I'm not advocating this viewpoint, but perhaps it isn't out of the question.

Alternatively, I suppose that one could think that schematic letters even in empirical theories are required to be indefinitely extensible (even to vocabulary employed only in extensions of the original theory that the advocate of the original theory rejects). That would involve giving up 2 (and the last sentence of 3.) On such a view, the claim of conservativeness might be deemed misleading: it would still be correct if $N+S$ is understood as simply the union of N and S, but on this viewpoint that might seem too anemic a reading of "adding" S to N. But I don't see that this viewpoint has much appeal. (Indeed, I'm inclined to stipulate that to regard a comprehension schema for regions as indefinitely extensible in this way *just is* to treat the schematic letters for predicates as logical variables, and thus to invoke at least a fragment of second-order logic (the Π_1^1 fragment). Similarly, I'm inclined to stipulate that to regard an induction schema for a finiteness predicate as indefinitely extensible *just is* to treat finiteness as a logical notion.)

0.4.4. *Summary: Conservativeness and Representation Theorems*

To summarize, I concede to the book's critics that Chapter 1, in combination with the middle chapters, was rather misleading: together, they could well be taken to suggest that it is possible to simultaneously maintain both the *syntactic* conservativeness of mathematics and the full representation theorems of those middle chapters. As Shapiro (1983a) rightly says, you can't have both.

I noted this myself in Chapter 9 of the book. I didn't there rule out keeping the full Hilbert-like representation theorems and taking the relevant conservativeness to be semantic conservativeness for the "complete logic of Goodmanian sums"; but I did make two not-very-developed alternative suggestions, one for going fully first order and the other for (in the terminology just used) going "almost first order" but using a primitive finiteness quantifier. In the book it probably appeared that this required giving up on representation theorems, but I made clear in the 1985a paper that this is not so, it merely requires generalizing them by allowing for representing fields that are not quite the real numbers but are extremely similar. In the purely first-order case, we need no notion of conservativeness beyond the syntactic, but a price is that the representing fields may be non-Archimedean. In the "almost first-order" case the representing fields are all Archimedean, but we require a notion of ω-conservativeness, which goes beyond the syntactic but isn't nearly as strong as the kind of semantic notion of conservativeness contemplated at some points of the book. I now think that both the purely first-order and "almost first-order" options are vastly more attractive than the option with the "complete logic of Goodmanian sums".

So, while the early part of the book is rather misleading, I still think that the basic line of Chapter 1 is right: the only obvious requirement on a good mathematics is that it be conservative, not that it be true.

0.5. Indispensability

Platonism is usually construed as requiring both the existence of mathematical entities and the objectivity of mathematics (in the sense described earlier in this Preface); and the "Quine-Putnam indispensability argument" (Putnam 1971) is usually taken as an argument for platonism in this strong sense. At least when platonism is so construed, then I stand by the view that there is no serious argument for it other than the indispensability of mathematical entities to things outside mathematics (though this could include more than basic physics, e.g. it could include the metatheory for logic). But how good is even that argument? Of course if the nominalization program of *SWN* could be carried out, it would not ultimately be a good argument for any form of platonism. But in this section I will take for granted the premise that the program of nominalizing theories is likely to fail. And my question is, how good is the argument for platonism on that assumption of indispensability?

We need to separate the question of how good the indispensability argument (taking that indispensability premise for granted) is as an argument for mathematical objects and how good it is as an argument for objectivity. The extensive recent literature is focused much more on the first than on the second,[F22] and I will start out with that.

[F22] An exception is Hellman 1989, who advocates using modality in connection with higher-order logic to eliminate mathematical objects from physical theory and elsewhere, but thinks that the indispensability argument is still important for objectivity. (Putnam's own view wasn't all that different: see Putnam 1967.) I'll say a bit about his view in section 0.7F.

0.5.1. Objects

I now think that the indispensability argument for the existence of mathematical objects is somewhat overblown. There's a great deal of recent literature in this direction (e.g. Yablo 2005 and 2012, Sober 1993, Melia 2000, Maddy 1992, Leng 2010 and 2012). There is also some interesting work in partial defense of the indispensability argument (e.g. Colyvan 2001, 2010, 2012, and Baker 2005); and one obvious challenge to those who would deny the relevance of the argument is that doing so seems to remove the means of deciding ontological questions. I don't have a worked out opinion on all of the issues involved in this literature, but I will make some tentative remarks on some of these issues and on some related ones.

First, on my understanding of the indispensability argument: discussion of this argument (especially by those who emphasize its roots in Quine as opposed to Putnam) is very often tied to some doctrine of "confirmational holism", whose content tends to be left unclear. (It is sometimes formulated as the claim that all parts of our best theories are equally confirmed; that formulation has the advantage of being fairly clear, but the disadvantage of being totally preposterous.[F23]) In stating the indispensability argument it's better to avoid any explicit talk of confirmation, and talk instead of what to believe. As a crude first try, we might say:

We should believe (e.g.) quantum electrodynamics, since it explains so many things and there's no decent competitor that does so. Quantum electrodynamics entails that there are mathematical entities. So we should believe that there are.

(A somewhat more nuanced version would take into account that there are bound to be competing theories we don't know about, and involve as a premise that the competitors are likely to also entail that there are mathematical entities.) I don't see that anything worth calling "confirmational holism" is implicit in this.

Quine advocated the indispensability argument in the context of the view that mathematics is fairly straightforwardly empirical—empirical in just the way that entrenched physical principles like the conservation

---

[F23] Does anyone really think that, in the early years of general relativity, the existence of gravitational waves and of black holes was as well-confirmed as the equivalence of inertial and gravitational mass or the gravitational redshift?

of energy-momentum are. (Some sort of "confirmational holism" may be relevant to *this*.) I was inclined to resist that view: indeed I took the conservativeness of mathematics, according to which good mathematics is compatible with any internally consistent theory of the physical world, to tell against the idea that mathematics has any empirical content in any ordinary sense.[F24] Quine's view seems to have been that if a certain mathematical theory S proves indispensable to (e.g.) a basic physical theory T, then the empirical evidence for T would be evidence for S, and so S would in a straightforward sense be empirically confirmed. But as Sober has emphasized, the assumption that evidence for T is evidence for S appears to rely on a discredited view of evidence and confirmation (the "consequence condition"). Indeed, my focus on conservativeness was supposed to indicate that S was already completely acceptable as mathematics, so no evidence could raise its status.

But there is another way of thinking about the bearing of indispensability arguments on empiricality, that doesn't rely on the consequence condition. For if we assume that there are possible physical theories in which mathematical entities are indispensable as well as possible ones where they aren't, the indispensability argument seems to require us to think that empirical evidence that favors theories of the first sort *over theories of the second* should enhance the extent to which we have literal *belief* in mathematical entities (as opposed to merely accepting them on conservativeness grounds).[F25] So if, for instance, classical physics were completely nominalizable while quantum physics isn't, then the early 20th century would have provided massive empirical evidence for the existence of mathematical objects (even though our mathematical theories would have already been *fully acceptable* for use in many contexts on conservativeness grounds). I don't think I really faced up to this unappealing consequence of taking indispensability arguments as

---

[F24] I'd occasionally flirted with, though never advocated, the view that *logic* is empirical. If it is, then since mathematics like every other discipline uses logic, maybe that's enough to make mathematics in some sense empirical. But I took the conservativeness of mathematics to show that, on any reasonable measure of "empirical content", mathematics has no empirical content beyond that of the logic it employs.

[F25] More carefully: the evidence favors theories of the first sort over "atheistic variants" of theories of the second sort: variants which dispense with the mathematical entities and add the claim that there are no such things. These atheistic variants have the same empirical contents as the originals.

decisive for mathematical ontology. My hope was that mathematical entities would prove indispensable to *any* reasonable fundamental physical theory, whether correct or not, so that insofar as evidence is limited to the selection of one "reasonable" fundamental theory over another it would prove irrelevant to the belief in mathematical entities. Of course this would make the task of establishing the requisite dispensability all the harder.

Indeed, to some extent my hope wasn't limited to fundamental theories. Of course, non-fundamental theories (including those in physics, such as thermodynamics and continuum mechanics) are accepted only as approximations, so any nominalistic formulation of them would obviously have to make use of false idealizing assumptions. That doesn't undermine the interest of trying to provide intrinsic nominalistic formulations of such theories based on such false assumptions. But it isn't clear that a reasonable response to an indispensability argument would *require* this: see the remarks on "intellectual doublethink" to come.

In any case, a view with obvious appeal is that indispensability arguments in mathematics simply have no force: that mathematics is just the framework in which physical theories (and other theories, including e.g. theories in metalogic) are stated, and that there's no reason to literally believe the framework. Since the physical theory entails aspects of the framework, then if we don't believe the framework we can't literally believe the theory, but can still believe it *relative to the framework*.

Despite its obvious appeal, this view worried me, because I didn't know how to respond in any detail to someone who took the analogous view of subatomic particles: that talk of them is just the framework in which modern physical theories are presented, and that there's no reason to literally believe in them. Of course I was aware of a possible response: that we should regard as framework only those aspects of a theory that *play no causal role*. But that seems unhelpful: a claim about a function from configuration space to the complex numbers does enter into explanations of how a physical system behaves, and if such a claim can't be expressed nominalistically then it's hard to deny that it plays a causal role (or at any rate, hard to see the content of such a denial). Why then shouldn't it be literally believed? Perhaps the claim would be that what are to be literally believed are only facts about causally relevant *entities*, and this excludes the complex numbers (and the configuration space, and the functions from the latter to the former). But even aside from general unclarity in the notion of causation, the notion of a causally relevant *entity* is *especially* unclear, and a platonist might well respond to the suggestion (as I did on behalf of the platonist in Field 1998b) by saying that

> if mathematical entities are theoretically indispensable parts of causal explanations ..., there seems to be an obvious sense in which they are causally involved in producing physical effects; the sense in which they are not causally involved would at least appear to need some explanation (preferably one that gives insight as to why it is reasonable to restrict [indispensability arguments] to entities that are "causally involved" in the posited sense). (p. 400)

Lacking a good response to the apparent analogy between mathematical objects and subatomic particles, I took a hard line: that it is "intellectual doublethink" to fully advocate a physical theory that postulates mathematical entities while at the same time denying the existence of mathematical entities. ('Fully advocate' means something like 'advocate as literal truth, not as mere approximation'. So the "intellectual doublethink" charge has no direct application to non-fundamental theories, which inherently involve approximations.) Of course one might without doublethink accept the physical theory as a temporary expedient that one would have to make do with until a program for eliminating the mathematical entities was carried out; and even after the "nominalization" had been carried out, one might without doublethink accept the original "un-nominalized" theory as a convenient calculational or heuristic device whose legitimacy turns on the nominalistic theory that underpins it.

Many of the opponents of indispensability arguments mentioned early in this section have stressed that in theories like continuum mechanics—non-fundamental theories which presumably are accepted only as approximations—we make a lot of use of knowingly false assumptions about the physical world. They seem to take this as showing that it's perfectly OK to use mathematical assumptions that we believe false in physical theories that are assumed to *precisely describe* the physical world. It's hard to see their argument here, and indeed the argument could well seem to go the other way. For a great deal of effort in physics is devoted to explaining, from fundamental theories assumed for the sake of argument to be precisely true, why the theory with false assumptions works as well as it does. So if we assume mathematics false, shouldn't the moral be that we must explain why theories that employ it work as well as they do, using a literally true theory in the explanation? That is the project of *SWN*, and it is still my view is that it is highly advantageous to do precisely that.[F26]

But the indispensability argument requires not only that it is *better* to explain the use of ideal or fictional objects in terms of things one literally believes in whenever one can, but that it is *compulsory*. This seems like dogma: why should we rule out that there might be circumstances where even in one's fundamental physics one has to employ some devices that one thinks of instrumentally? Yablo 2005 raised this challenge forcefully: once one has granted (as I did in my discussion of conservativeness and representation theorems) that it is legitimate and useful to appeal to fictitious objects to facilitate the adjudication of inferences, why shouldn't one also grant that it can sometimes be legitimate and useful to appeal to such objects directly in the representation of the physical world (including the representation of features that enter into explanations)? Yablo didn't stress to the extent I would have that it is *better* to look for theories and explanations that don't involve fictions, when they are available, but still, his challenge seems to me correct. If so, the "intellectual doublethink" charge was wrong.

This does make pressing the question of how to distinguish between the instrumental part and the non-instrumental. I don't pretend to have anything like a general answer. In the special case of mathematics, a natural idea (somewhat motivated by the remarks on cardinality quantifiers above, and perhaps also by recent literature on deflationary theories of truth[F27]) is that without ontological commitment, mathematics can be used both to increase our power to describe the physical world and to make generalizations about it. As Yablo puts it, "The nominalist rejects mathematical *ontology*, not mathemathical *typology*" (2012: 1023). The restriction of mathematics to this expressive and generalizing role *limits* the way that mathematics can be used in explanations, but certainly doesn't *eliminate* its use in explanations: for instance, mathematics might provide the most efficient way to state what is in common to all the stable states of a certain physical system, or what oscillating pendulums and electronic oscillator circuits have in common. By contrast, the role of subatomic particles in physics seems to have more than such a purely expressive/generalizing role.

[F26] This paragraph has been influenced by discussion with Cian Dorr.

[F27] An analogy to deflationary views of truth is pressed in pp. 94–5 and note 16 of Yablo 2005.

Of course, these remarks are extraordinarily sketchy: we need a much more thorough investigation of the distinction between allowable and unallowable ways for a nominalist to use mathematics in explanation. (Section 7 of Yablo 2012 gives some first steps.) And I'm not sure to what extent quantification over mathematical entities is really required in these cases: for instance, the harmonic oscillator equations in a given domain can certainly be presented nominalistically, and their common form in various domains should be evident without the mathematical entities. Still, I'm convinced that we should allow for the possibility of taking some features of even fundamental theories as instrumental (though we should also try to minimize the extent to which we rely on instrumental devices). As a consequence, I'm still inclined to the anti-platonistic metaphysics of the book, despite skepticism that the project of nominalistically formulating every fundamental theory we care about can be carried out.

But I don't think that this skepticism completely undermines the relevance of *SWN* to the issue of nominalism, or anti-platonism more generally. For one thing, I think the book helped clarify the role that mathematical entities play in science, and how this role differs from the role that physical entities play. It was already clear *that* the roles differ: as noted, it's natural to say that mathematical entities aren't causally involved in producing physical effects in the way that electrons and quarks are. But as also remarked, it is less than clear what this difference comes to, given that mathematical entities like physical entities are appealed to in typical (i.e. platonistic) explanations of those physical effects. I think it likely that the content of the difference in the "causal role of the entities" must be based on differences in the roles of these entities in explanations that can be expressed independently of causal talk. And I think that *SWN* went at least *some* way toward clarifying that difference in explanatory role. In part it did this by making plausible that *much* of the role of mathematical entities in science is to serve in representations that are not needed in the underlying ("intrinsic") formulations of the physical theory and that have a systematic arbitrariness. (Of course there were antecedents to this, both in the literature on measurement theory and the literature on tensor formulations of physical theory; but *SWN* carried the ideas further.)

SWN was almost certainly over-optimistic in suggesting that, at least in fundamental science, all use of mathematical entities in representing non-mathematical structure is eliminable. But even if there are theories that can't be presented without the "fiction" of mathematical objects, I think that the methods of the book could be used to express a great deal of their content in a nominalistic manner, far more than had hitherto been believed. I think that in evaluating the claim that a certain representational device in a theory should be regarded as a fiction, it matters *how much* can be done without what is claimed to be a fiction. The exact role that these devices are playing in the theory also matters: an analysis of that role may lessen the extent to which an argument for the non-fictional status of these entities is convincing. For instance, if the role of sets in physical theory were simply to allow us to assert the local compactness of physical space, would this really provide a convincing case for platonism?

I recognize of course that much of what I've said here (especially the last point) is vague and deserves more discussion; let me simply say that the proper context for the debate referred to, between Colyvan and Baker on the one side and Yablo, Sober, Melia, Maddy, Leng, and others on the other, is best carried out in the context of an appreciation that the essential role of mathematical entities in formulating scientific theories (especially fundamental scientific theories) is far less than it initially appears to be.

0.5.2. Objectivity

I've been speaking of the indispensability argument as an argument for *the existence of mathematical objects*. But as I've mentioned, platonism is often understood as requiring in addition (or instead) a kind of *objectivity* of mathematics, exemplified by the view that questions such as the size of the continuum must have uniquely correct answers. It seems to me pretty clear that most indispensability arguments will not secure this sort of objectivity. For the most convincing indispensability arguments are arguments that we can't adequately characterize physical structures except by reference to mathematical structures: e.g. that we can't adequately characterize quantum systems without configuration spaces, or gauge fields without fibre bundles, or whatever. In these examples, the role of mathematical entities in theorizing that is not easily shown indispensable is their role as *exemplars of possibilities*: mathematics provides rich structures that are not found in the physical world but that are nonetheless highly useful in describing the physical world since we take the physical world to contain isomorphic images of substructures of them. (See Shapiro 1983b.) But in its role as a source of rich structures, set theory with one choice of continuum size and set theory with another choice are equal: if mathematics with one choice for the size of the continuum were used in an application, one could use mathematics with another choice for the size as well (if need be, by constructing a model for the second mathematics within the first).[F28] Even supposing that it is an objective matter whether physical lines contain $\aleph_{23}$ points (which of course could well be doubted), it doesn't follow that the acceptability of a mathematics that says that the continuum has size $\aleph_{23}$ should turn on this: for on a platonist view, physical lines are one thing and the mathematical continuum (the set of real numbers as defined via Dedekind cuts) another. (That had better be the platonist view, since space-time is just one application of the real numbers, and there's no obvious reason to think that in other applications, e.g. to degrees of belief, the non-mathematical reality to which the reals are typically applied has the same structure as the lines of space-time.) Whatever one's view of the size of the platonist continuum, physical lines can be accommodated, and so it isn't obvious how an indispensability argument could motivate the objectivity of mathematics.

0.6. Other Forms of Anti-Platonism

The book may have implicitly underestimated how many ways there are of being "anti-platonist" in an intuitive sense without saying that mathematical theories aren't literally true if taken at face value. One apparent alternative that I'm attracted to would be to regard mathematical theories as "true by convention": any mathematical theory that is conservative is true for any community that adopts it.[F29] (If one were to defend this line, one would need to decide between regarding mathematical entities as existing by convention, or denying that 'p' being true by convention entails p. The first alternative strikes me as more attractive—though the choice between them might itself be a matter of convention. Either way, one needs a bit of care to avoid anomalous behavior in counterfactuals whose antecedents concern our having

---

[F28] The limitations on "inner model" proofs do not prevent this: the second model could, for instance, be the result of collapsing the Boolean valued model given in an independence proof by an ultrafilter.

[F29] A variant, perhaps cleaner, would be that if there is a possible convention according to which a given mathematical theory would be true by convention, then it is true of part of mathematical reality already, even if the convention isn't adopted. This would be pretty close to Mark Balaguer's "plenitudinous platonism" (1998), which I think shares a key feature with the anti-platonism of *SWN*: it rejects the idea that on questions such as the cardinality of the continuum there is a uniquely right answer.

adopted conventions that conflict with those we've actually adopted; perhaps a two-dimensional modal framework is the way to go.)

Quine 1936 had given a famous regress argument against the view that what we regard as *logical* truths get their truth by convention (and presumably, that what we regard as truth-preserving logical inferences get their truth-preservingness by convention). It is most directly an argument against their being true (or truth-preserving) by *explicit* convention; but it is widely thought (rightly or wrongly) that his argument extends to implicit convention too, on the grounds that one would need logic to determine what the consequences are of consistently employing the basic rules that describe the implicit convention. If one assumes that logic and mathematics have the same status, then his arguments would indirectly tell against truth by convention for mathematics too. But in the book I denied that logic and mathematics have the same status, so the idea that mathematical theories are true by convention would seem an option even if one concedes that Quine's case against the conventionality of logic extends to implicit convention.[F30]

I call this an *apparent* alternative, since there doesn't seem to be a whole lot of difference between regarding a theory as true *by the convention we've adopted* and its being true *according to the fiction we've adopted*. But perhaps "truth by convention" can more easily be regarded as a kind of literal truth, and perhaps there might seem to be less force in a demand to avoid "truth by convention" in basic physics than to avoid "fictional truth" in basic physics.

There may also be other ways of developing the idea that mathematics is true, but not in a way that would support "real platonism". But I remain less than convinced that the distinction between any such "truth but of a lesser sort" view and a "not literally true" view is of any deep significance.

0.7. Miscellaneous Technicalia

(A) Integral Calculus

Although this has been little commented on, *SWN* was incomplete even for Newtonian gravitation theory, because (as I noted in its "Preliminary

---

[F30] The admirable discussion of mathematical epistemology in Rosen 2001 seems to me to point in the direction of mathematical conventionalism, though Rosen does not draw that conclusion.

Remarks") it didn't deal with integral calculus, only differential. While I believed, and still do, that this could have been remedied had I taken the trouble to do so, I don't think I fully appreciated the extent to which we don't have a satisfactory formulation of the theory until it is remedied.

More fully, the nominalistic theory that I give generates a platonistic theory that doesn't explicitly talk about masses, but only about mass-densities at points. Of course, the mass-densities at points in conjunction with the volume element determined by the metric are enough to determine the masses assigned to regions (i.e. the masses of what occupies those regions), when the mereological space is atomic as *SWN* assumes. But this supervenience of the masses on the mass-densities (plus metric) doesn't seem to me enough: after all, our observations are of (comparisons of) the masses of extended objects, not of the mass-densities of points, and we ought to want a theory in which we can express our observations. If we keep comparisons of mass-density as primitive, we need an explicit theory of how they, together with the geometric properties, determine the comparisons of mass. That's what a nominalistic analog of the theory of (multi-dimensional) integration would provide. (We could also switch to using mass-comparisons rather than mass-density comparisons as primitive; that's probably more natural. But even so, we'd need to then introduce mass-density comparisons derivatively if we are to state the field-theoretic form of the Newtonian theory, and the integration theory would be needed for a full picture of how the masses and mass-densities, or even volumes and distances, interrelate.)

A somewhat related point is that the treatment of Laplaceans is inelegant, and arguably not in accordance with the rather vague requirement of intrinsicness I had imposed.[F31] A better treatment would involve a nominalistic analog of the volume element 3-form, which is one (small) part of a proper treatment of multi-dimensional integration.

[F31] The systems of three vectors, orthogonal and of the same length, look very much like Cartesian coordinate systems (though admittedly, in making the comparisons of the Laplacean at different points I didn't demand that the three directions be the same at the different points). What's relevant isn't the systems of three vectors, but the volumes they span; or rather, the comparison of volumes at different locations. (The gradient contravector is also required in a proper treatment, which means that the material in section 8H and 8I would need to have come earlier. The order in the text is rather odd anyway, since of course the Laplacean is just the divergence of the gradient.)

While I haven't worked out the details of a nominalistic counterpart to the theory of measure and integration, I don't have much doubt that the general idea of what I was doing could have been extended to deal with it.

(B) Point Particles

Another lacuna, noted and discussed on p. 223 of Arntzenius and Dorr 2012, is that I speak of point-particles in a context where there is a continuous mass-density distribution; and it is not obvious how to fix this. I appreciate the problem, but I think that it isn't specific to the nominalist project, it's an issue for the platonist theory being nominalized. (And while I could have considered action-at-a-distance formulations of Newtonian gravitation that don't raise the problem, I was trying to illustrate how field theories more generally would be dealt with; and the problem is a general one for standard platonist formulations of field theories.)

(C) The Gravitational Constant, and Comparativism about Quantities

There is another possible lacuna whose significance was only recently brought home to me, by David Baker (manuscript (a)): my nominalization of Poisson's equation is really just a nominalization of the claim that *there is* a positive value of the gravitational constant G for which the equation holds; it doesn't fix the actual value of G.

Here's how I defended this:

> The absolute value of the proportionality constant has no invariant significance within this theory: to give it significance you have to impose independent constraints on the mass scale and the scales of other quantities. (p. 79)

That is literally true, but may obscure the fact that the proportionality constant *does* have significance *relative to* a choice of scale for the other quantities. As Baker emphasizes, this is important because the value of the gravitational constant is relevant to whether the gravitational binding energy of a system of moving particles suffices to hold it together.

So something needs to be said about how that relative value of the gravitational constant is to be stated in nominalistic terms. Of course it's easy enough to do that; the simplest way is to fix three points in a region of varying gravitational potential on the trajectory of a single specific particle (with non-zero mass). But there may seem to be something inelegant in focusing on three specific points of a given trajectory.

The issue here—and the focus of Baker's discussion—isn't really nominalism *per se*, but *comparativism*, the view that what's physically fundamental is scale-independent relations among objects. (Comparativism is defended in a platonistic setting in Dasgupta 2013, and that paper is Baker's focus.) But comparativism is fairly central to *Science Without Numbers*,[F32] so the issue deserves discussion.

An initial worry about comparativism is that the mass scale, like the gravitational constant, seems physically relevant: uniformly increasing all masses with the gravitational constant fixed makes for more gravitational collapse, in just the way that increasing the gravitational constant with the values of the masses fixed does. So it is natural to accept counterfactuals like

If the positions and velocities of all particles relative to an inertial frame at a given time were just as they actually are, and so were their mass ratios, but the absolute values of the masses were bigger, then (with the gravitational constant held fixed) there would be more gravitational collapse.

Since increasing all masses by the same factor is like increasing the gravitational constant, the discussion of how the value of that is fixed might suggest that we simply modify the counterfactual by replacing "but the absolute values of the masses were bigger" by "but the angle at the temporally second vertex of the triangle determined by three points on the trajectory of such and such particle was bigger". (We could then drop the parenthetical clause about the gravitational constant.) But in the context of such a counterfactual it may seem more obviously inelegant to rely on the specific points on the trajectory of a specific particle. (We could instead replace the phrase by "but the trajectories of *all* particles were more concave", but that not only invokes a great deal of redundant information, it tends toward trivializing the counterfactual.) That's the initial worry.

But in fact there's no need to deal with the counterfactual in such an inelegant way. For as Baker pretty much notes,[F33] the language of *Science Without Numbers* will require talk of the velocity of a particle (relative to a rest frame) at a given time to be restated, in terms of position (relative

---

[F32] Dropping it would pretty much require moving to a theory with primitive mass properties, of the sort mentioned early in Chapter 7 and developed by Mundy.

[F33] Since his discussion isn't focused particularly on nominalistic views, I'm abstracting a bit.

to an inertial line) throughout some interval; and that is enough to determine the acceleration of all particles throughout that interval. In other words, the counterfactual would have to read something like

If the positions of all particles (relative to an inertial line) *throughout a given interval* were just as they actually are, and so were their mass ratios, but the masses were bigger, then (with the gravitational constant held fixed) there would be more gravitational collapse.

But then the antecedent of the counterfactual is a nomic impossibility: even in the platonistic theory, the rest of the antecedent together with the fixed gravitational constant rules out the masses being bigger.

Baker's paper, while dismissing the initial worry in pretty much this way, argues that a more sophisticated worry akin to it survives. This is based on toy examples according to which there are periods of time during which all accelerating particles are in regions of unchanging gravitational potential, so that there is nothing during these periods to fix the value of the gravitational constant. Baker then points out that, if these idealized examples are accepted, there will be odd violations of determinism and of certain natural locality principles in theories that take mass-comparisons as opposed to quantitative mass as basic. While not crippling, this looks on its face as something of a worry for *SWN*, or any other view which eschews the use of scale-dependent relations between physical objects and numbers in the physical laws.

One thing to note about the threat is that Baker's examples are toy not only in employing universes much simpler than the actual, but in making various not-obviously-innocent idealizations. His most convincing example is his "shell-plus-projectile world", with a uniform spherical shell and a projectile that somehow can pass through it; while the particle is in the interior of the sphere there is no net gravitational force on it, so that the path of the projectile during the interval when it's inside the sphere doesn't determine the gravitational constant. Baker needs that *nothing* during that interval that can be stated in non-quantitative terms determines the value of the constant; for this it is essential that the shell be of continuous matter whose parts don't accelerate in the slightest. (No electrons circling around nuclei, etc.) And that requires that there be non-gravitational constraint forces among the parts of the shell that precisely counteract the gravitational forces on them, *even though those gravitational forces on the parts are changing slightly as the projectile moves from one part of the interior to another*. It would be a tall order to invent laws for the constraint forces with this feature.

Nonetheless these are the kind of toy examples one often uses in physics, and I'm not entirely comfortable dismissing them as irrelevant. Fortunately there are other resolutions of the "sophisticated worry" from two paragraphs back. In particular, Baker has another paper (manuscript (b)) with an alternative solution, which invokes the sort of "mixed comparative relations" that I will motivate on two independent grounds, in subsections (D) and (E).

(D) Avoiding the Notion of Finiteness

In section B of Chapter 8, I said that to define a predicate comparing ratios of distances with *ratios of scalar intervals*, we need to invoke the binary predicate of one region 'containing fewer points than' another, or a restricted version of this that applies only between finite regions. This raised an issue for the purely first-order version of the theory: in absence of representation theorems that restrict distances and the scalar in question to Archimedean scales, there will be models in which the ratio-comparison predicate has non-standard application even to finite regions. I argued that this isn't a decisive difficulty, because we could still get a pretty good axiomatization of the predicate using an induction schema; still, it provided some push toward a not-purely-first-order theory with a primitive cardinality quantifier.

Burgess and Rosen 1997: 191 suggest that this was all beside the point, since for defining a predicate comparing ratios of distances with *other ratios of distances* we don't need cardinality comparisons, we can use the multi-dimensional affine structure. But while that's fine for comparing distance-ratios to *other distance ratios* (at least in the case of flat infinite space), it doesn't deal with the mixed-scale predicates that were at issue in section 8(B).

Burgess 1984: 393 makes a different suggestion (which he credits to Saul Kripke) that does deal with such predicates: that we continue to use a 'contains fewer points than' predicate but define it, using regions consisting of non-intersecting physical arcs to express 1–1 correspondences. While I don't find the suggestion altogether appealing, it would seem adequate to the needs of the theory. Of course, axioms about these physical arcs would have to be added to ensure that the needed properties of 'fewer than' followed from the definition. I'd suspect that taking 'contains fewer points than' as primitive and axiomatizing it gives rise to a simpler theory, but confess that I haven't thought about this enough to say so with confidence.

A quite different alternative, which avoids the need for cardinality comparison in the theory, is to give up on *defining* the mixed-scale predicates in section 8(B), and instead altering the basic ideology to include some of them. Indeed, Burgess himself suggests this earlier in the paper, with his Q_s predicate (Burgess 1984: 390–1). One would need only one such primitive Q_s predicate for each scalar in the theory: it relates the ratios of that scalar to geometric ratios, and the affine structure can then do all the rest. (We could drop the betweenness and congruence predicates that I used for the scalar; they would be definable from the Q_s predicate together with the geometric predicates.) As Cian Dorr emphasized to me, this approach avoids not only the need for cardinality comparisons but also the need for regions bigger than a point (for anything other than ensuring that the lines have structure more like the real numbers): in other words, it allows for a nominalistic theory that is elementary in the sense of Tarski 1959. Moreover, as we'll see next, there are other reasons for using such mixed-scale comparative predicates. All in all then, this approach probably yields the simplest overall theory.

But I remind the reader of the claims in section 0.1: there are reasons independent of physics for wanting 'fewer than', 'finitely many', and other cardinality quantifiers that aren't definable in first-order logic. Such quantifiers have more general applications than in comparing the number of points in a region: e.g. for any predicate F of regions, it would allow us to say that there are only finitely many regions that are F. With a bit more apparatus, e.g. predicative second-order logic, this would enable you to define a fair notion of compact region. While I didn't need this for my purposes in the book, I'm inclined to regard it as perfectly legitimate even from a nominalistic point of view that eschews fictional devices.

(E) Point Masses without Restrictive Structural Assumptions

One of the reasons I assumed an absolutely continuous mass distribution rather than discrete point masses is that the obvious representation theorems for mass require structural assumptions about the quantity that don't obviously have to be met for point masses. For instance, given point masses, there's no obvious reason to assume one for which all other point masses are multiples of it in mass; but without that, it seems unobvious how to give a representation theorem. Arntzenius and Dorr point out that for certain theories there are tricks one can employ for doing this, but conclude that a general approach requires positing a special "mass space"; similarly, a special "charge space", and so on for other quantities. (As they note, this is really the same as the proposal of Mundy 1987, of taking the basic comparative notions employed in the theory to relate physical properties such as that of having a particular mass; they argue (p. 229) that the Mundy formulation may disguise the commitments of the view.)

Zee Perry 2015 has argued, though, that there is an attractive way (which she motivates on independent metaphysical grounds) to handle this without quantifying over physical properties or points of quantity spaces. The idea is to take quantities like mass and charge as derivative on space-time, by axiomatizing them with "mixed" comparative predicates (of objects or points of ordinary space) that involve distances as well as the quantities of direct concern. The predicates in question will be like Burgess's Q_s, in effect comparing ratios of the quantities in question (e.g. mass) with ratios of distances (or of volumes, or of some such spatio-temporal magnitudes). Since the distances (or volumes or whatever) can vary continuously even if the masses don't, this handles the worry rather well, and I'm inclined to think this a considerable improvement over the approach in the book.[F34] Especially since it has the additional advantages noted under (D).

And as Baker (manuscript (b)) notes, a particular version of this also resolves the issues in (C) in a way that doesn't require kvetching about the toy examples. The key is to use, instead of ordinary mass-comparisons, predicates that compare mass multiplied by the square of time-interval divided by the cube of distance, or equivalently, density multiplied by the square of time-interval. This in effect takes mass just to *be* the gravitational acceleration it induces at a distance r multiplied by the square of that distance; there is no need of the constant G to get from one to the other, so the issues raised by the inelegant fixing of it disappear.

(F) Dispensability via Modality?

In some papers in the 1980s, in particular in Field 1988, I took a skeptical line toward the idea that there is an interesting way to use modality to

---

[F34] These new primitives can be viewed as comparisons of single quantities, just less familiar ones than usual: in the volume case, we're comparing the average mass-density of regions, in the distance case it's a comparison of what we might call "matter-displacement" (mass times distance displaced).

dispense with mathematical objects. Part of my criticism was directed in particular toward nominalistic theories of the sort

$T^\lozenge$: Possibly, the concrete world is just as it in fact is, and T.

In particular, I argued that theories of this sort are "too cheap" in that they could be trivially modified to avoid commitment to subatomic particles. But as Dorr 2010 argues, while this cheapness argument is telling against theories of the sort $T^\lozenge$, it isn't clear that they have force against an alternative style of modalized theories:

$T^\square$: Necessarily, if M and the concrete realm is just as it in fact is, then T

(where M is an impure mathematical theory). And it is theories in the style of $T^\square$ rather than $T^\lozenge$ that were advocated shortly afterward in Hellman 1989.

But the main points of Field 1988 apply at least as well to $T^\square$ theories as to $T^\lozenge$. These points concern the question of how we are to spell out the content of the concreteness conjunct ("the concrete realm is just as it in fact is"), without appeal to mathematical entities. After all, we normally describe the physical world using mathematical entities; but obviously if the claims of form "A if and only if actually A" that are built into the concreteness conjunct were taken to include those where A quantifies over mathematical entities, the theory would in no way be nominalistic. On the other hand, if such platonistic A are excluded, there is a substantial worry that we won't have enough left to sufficiently describe the concrete world.

This worry is actually worse for $T^\square$ theories than for $T^\lozenge$ theories, because for a $T^\square$ theory to be true the description of the concrete world must be rich enough so that it together with M necessitate the platonistic theory. Just how serious the worry is may depend on how austere our conception of necessity is. (I myself am rather puritan about this, being suspicious of modal talk that isn't explainable in terms of *logical* modality, where the logic is first order, or first order plus some cardinality quantifiers. But that's too big an issue to discuss here.)

It is of course clear that, were the non-modal nominalization program successful, one could use the nominalistic theory N to formulate enough of the concreteness conjunct to guarantee $T^\square$: replace that conjunct by "N if and only if actually N". But it is equally clear that that would make the use of modality redundant. Whether there is any satisfactory way short of this to spell out the concreteness conjunct and the modal operator so as to guarantee $T^\square$ is a question I will not pursue here.

(G) General Relativity

Finally, I'll expand a bit on the suggestion in the book (last paragraph of footnote 44) for how to extend the nominalistic treatment to the curved space-time of general relativity. (This is still only the barest of sketches.)

One way to implement the suggestion would be to replace the betweenness and congruence predicates by more general primitives: perhaps a four-place predicate "point y is between points x and z relative to region R", meaning intuitively that there is a geodesic segment in region R with endpoints x and z that contains y; and a five-place predicate "x,y parallel-congruent to z,w along geodesics within R", meaning intuitively that any R-geodesic segment from x to y, when parallel transported along any R-geodesic segment from x to z, results in an R-geodesic segment from z to w. Using the first, we can define the idea of a region R being convex normal, meaning that (it has at least two points and) any two points in it are connected by exactly one geodesic lying wholly within it. (This implies that geodesics don't self-intersect in R.) Among the axioms governing this predicate will be that each point is contained in such a convex normal neighborhood. The second predicate is needed to parameterize geodesic segments: we parameterize such a segment, up to a scale factor, using parallel transport of short subsegments of it along it.

When R is convex normal, then for any x, y, and z there will be at most one w such that x,y is congruent to z,w along geodesics within R. There needn't be one: the convex neighborhood might be "too small near z". But we'll impose axioms that guarantee (i) that for each x, y, and z, some subsegment x,y' of x,y starting at x will be transportable along x,z; and (ii) that if there is such a w, then transporting any "shortening of x,y by scale factor α" along x,z will result in a "shortening of z,w by scale factor α" (where the scale factors are with respect to the parameters of the respective geodesics). Together, these mean that the information about parallel transport of vectors in the tangent space that might appear to be lost in the translation to geodesic segments of limited length isn't really lost. (This is just another example of a technique employed several times in Chapter 8, of using the linearity of derivatives to overcome limitations in the range of quantities.)

This approach works for affine-connected manifolds generally; no metric is required. Similarly, the suggestion for how to treat an analog of particular covariant tensor fields on the manifold doesn't make use of a metric. (Of course it isn't an exact analog, because though a representation theorem for tensor fields will assign a real number to sufficiently small geodesic segments emanating from a point, it will do so only in a way that depends on a scale for the tensor-like quantity that's in question.) In manifolds that do have a metric, the metric is treated (as in platonistic theories) as just another covariant tensor field (with a requirement that it be compatible with the affine connection, in the sense that the metric applied to two geodesic segments emanating from a point has the same value as the metric applied to the result of transporting those vectors along a geodesic to a different point).

I'm not sure whether it's possible, without relying on a metric, to give an attractive extension of this treatment to tensors that are contravariant in some indices; but I think any such extension would be more complex. In the footnote I didn't pursue this, because the concern there was with general relativity, where there is a metric; it allows us to translate between contravariant and covariant indices.[F35]

[F35] Thanks to Cian Dorr, Marko Malink, Chris Scambler, Trevor Teitel, and Dan Waxman for helpful comments on an earlier draft of this Preface.

Bibliography for Second Preface

[1] Arntzenius, Frank and Dorr, Cian 2012. "Calculus as Geometry". In Arntzenius, *Space, Time and Stuff*. Oxford: Oxford University Press, pp. 213–78.
[2] Baker, Alan 2005. "Are there Genuine Mathematical Explanations of Physical Phenomena". *Mind* 114: 223–38.
[3] Baker, David (manuscript (a)). "Some Consequences of Physics for the Comparative Metaphysics of Quantity" (May 2013).
[4] Baker, David (manuscript (b)). "Comparativism with Mixed Relations" (April 2014).
[5] Balaguer, Mark 1998. *Platonism and Anti-Platonism in Mathematics*. Oxford: Oxford University Press.
[6] Burgess, John 1983. "Why I am Not a Nominalist". *Notre Dame Journal of Formal Logic* 24: 93–105.
[7] Burgess, John 1984. "Synthetic Mechanics". *Journal of Philosophical Logic* 13: 379–95.
[8] Burgess, John 1991. "Synthetic Mechanics Revisited". *Journal of Philosophical Logic* 20: 121–30.
[9] Burgess, John and Rosen, Gideon 1997. *A Subject with No Object: Strategies for Nominalistic Reinterpretation of Mathematics*. Oxford: Oxford University Press.
[10] Carnap, Rudolph 1950. "Empiricism, Semantics and Ontology". In his *Meaning and Necessity*. Chicago: University of Chicago Press, 1956, pp. 205–21.
[11] Colyvan, Mark 2001. *The Indispensability of Mathematics*. Oxford: Oxford University Press.
[12] Colyvan, Mark 2010. "There is No Easy Road to Nominalism". *Mind* 119: 285–306.
[13] Colyvan, Mark 2012. "Road Work Ahead: Heavy Machinery on the Easy Road". *Mind* 121: 1031–46.
[14] Dasgupta, Shamik 2013. "Absolutism vs. Comparativism about Quantity". In Karen Bennett and Dean Zimmerman (eds), *Oxford Studies in Metaphysics*. Oxford: Oxford University Press, vol. 8, pp. 105–48.
[15] Dorr, Cian 2010. "Of Numbers and Electrons". *Proceedings of the Aristotelian Society* 110: 133–81.
[16] Dummett, Michael 1994. "What is Mathematics about?". In A. George (ed.), *Mathematics and Mind*. Oxford: Oxford University Press, pp. 429–45.
[17] Dürr, Detlef and Teufel, Stefan 2009. *Bohmian Mechanics: The Physics and Mathematics of Quantum Theory*. Dordrecht: Springer.

BIBLIOGRAPHY FOR SECOND PREFACE

[18] Field, Hartry 1984. "Is Mathematical Knowledge Just Logical Knowledge?" In Field 1989/91, pp. 79–124. (The '91 edition has some corrections and a new postscript.)
[19] Field, Hartry 1985a. "On Conservativeness and Incompleteness". In Field 1989/91, pp. 125–46.
[20] Field, Hartry 1985b. "Can We Dispense with Space-Time". In Field 1989/91, pp. 171–226.
[21] Field, Hartry 1988. "Realism, Mathematics and Modality". In Field 1989/91, pp. 227–81.
[22] Field, Hartry 1989/1991. *Realism, Mathematics and Modality*. Oxford: Blackwell, 1989; revised edn Blackwell 1991.
[23] Field, Hartry 1991. "Metalogic and Modality". *Philosophical Studies* 62: 1–22.
[24] Field, Hartry 1992. "A Nominalistic Proof of the Conservativeness of Set Theory". *Journal of Philosophical Logic* 21: 111–23.
[25] Field, Hartry 1994. "Are our Logical and Mathematical Concepts Highly Indeterminate?". *Midwest Studies in Philosophy* 19: 391–429.
[26] Field, Hartry 1998a. "Which Undecidable Mathematical Sentences have Determinate Truth Values?". In H. G. Dales and G. Oliveri (eds), *Truth in Mathematics*. Oxford: Oxford University Press, pp. 291–310.
[27] Field, Hartry 1998b. "Mathematical Objectivity and Mathematical Objects". In S. Laurence and C. Macdonald (eds), *Contemporary Readings in the Foundations of Metaphysics*. Oxford: Blackwell.
[28] Field, Hartry 2008. *Saving Truth from Paradox*. Oxford: Oxford University Press.
[29] Hellman, Geoffrey 1989. *Mathematics Without Numbers*. Oxford: Oxford University Press.
[30] Hodes, Harold 1984. "Logicism and the Ontological Commitments of Arithmetic". *Journal of Philosophy* 81: 123–49.
[31] Kripke, Saul 1975. "Outline of a Theory of Truth". *Journal of Philosophy* 72: 690–716.
[32] Leng, Mary 2010. *Mathematics and Reality*. Oxford: Oxford University Press.
[33] Leng, Mary 2012. "Taking it Easy: A Response to Colyvan". *Mind* 121: 983–95.
[34] Maddy, Penelope 1992. "Indispensability and Practice". *Journal of Philosophy* 89: 275–89.
[35] Malament, David 1982. "Review of Hartry Field, *Science Without Numbers*". *Journal of Philosophy* 79: 523–34.
[36] Melia, Joseph 2005. "Weaseling Away the Indispensability Argument". *Mind* 109: 455–79.
[37] Meyer, Glen 2009. "Extending Hartry Field's Instrumental Account of Applied Mathematics to Statistical Mechanics". *Philosophia Mathematica* 17: 273–312.

[38] Mundy, Brent 1987. "The Metaphysics of Quantity". *Philosophical Studies* 51: 29–54.
[39] Perry, Zee 2015. "Physical Quantities: Mereology and Dynamics". Ph.D. dissertation, New York University (in progress).
[40] Putnam, Hilary 1967. "Mathematics Without Foundations". *Journal of Philosophy* 64: 5–22.
[41] Putnam, Hilary 1971. *Philosophy of Logic*. New York: Harper.
[42] Putnam, Hilary 1980. "Models and Reality". *Journal of Symbolic Logic* 45: 464–82.
[43] Quine, W. V. O. 1936. "Truth by Convention". Reprinted in his *Ways of Paradox*. New York: Random House, 1966, pp. 70–99.
[44] Quine, W. V. O. 1948. "On What There Is". Reprinted in his *From a Logical Point of View*. New York: Harper & Row, 1953, pp. 1–19.
[45] Rayo, Agustin and Yablo, Stephen 2001. "Nominalism through Denominalization". *Nous* 35: 74–92.
[46] Rosen, Gideon 2001. "Nominalism, Naturalism, Epistemic Relativism". *Philosophical Perspectives* 15: 69–91.
[47] Shapiro, Stewart 1983a. "Conservativeness and Incompleteness". *Journal of Philosophy* 80: 521–31.
[48] Shapiro, Stewart 1983b. "Mathematics and Reality". *Philosophy of Science* 50: 523–48.
[49] Simpson, Stephen 1999. *Subsystems of Second Order Arithmetic*. Berlin: Springer.
[50] Sober, Elliot 1993. "Mathematics and Indispensability". *Philosophical Review* 102: 35–57.
[51] Tarski, Alfred 1959. "What is Elementary Geometry?" In L. Henkin, P. Suppes and A. Tarski, *The Axiomatic Method*. Amsterdam: North-Holland, pp. 16–29.
[52] Yablo, Stephen 2005. "The Myth of the Seven". In Mark Kalderon, *Fictionalism in Metaphysics*. Oxford: Oxford University Press, pp. 88–115.
[53] Yablo, Stephen 2012. "Explanation, Extrapolation and Existence". *Mind* 121: 1007–29.

Note

I include, for historical interest, a letter from W. V. Quine, giving his initial reaction to the book (adjusted, no doubt, by politeness to the author who had sent him an unsolicited free copy).

I should add that his quasi-positive reaction didn't seem to last: to my knowledge he never in print indicated that the book made even the slightest dent in his confidence in platonism.

HARVARD UNIVERSITY

DEPARTMENT OF PHILOSOPHY
EMERSON HALL

CAMBRIDGE, MASSACHUSETTS 02138
(617) 495- 3913

April 23, 1980

Professor Hartry Field
School of Philosophy
University of Southern California
Los Angeles, CA 90007

Dear Hartry:

Your book came yesterday morning and I spent the day at it. It is an impressive piece of work: reasonable, ingenious, learned, and as central philosophically as can be. Moreover it appeals to my predilections, for, as you must know, I am a nominalist manqué from away back, and a reluctant platonist only in honest recognition of what have seemed to be the demands of science.

What to count as nominalism is a question of no great importance, though I shall get back to it. More important is the ontological economy and relative homogeneity that you achieve, whatever one's views of the objects that you keep. More important still, perhaps, is the economy of theory that you gain by what you call intrinsic formulation; namely, the resolving out of conventional units and measures in favor of the objective invariances that underlie the quantitative laws.

You seem to worry about your retention of mereology, or the theory of "Goodmanian sums," as a logico-mathematical vestige beyond recursive enumerability. I see no cause for misgivings here, beyond what one might feel regarding the space-time regions, nor any sense in which mereology is more distinctively logico-mathematical than your physical theory of space-time.

In our little "Steps toward a constructive nominalism" of long ago, Goodman and I welcomed mereology of physical objects, but denied ourselves space-time regions and all postulation of infinity. Very well: it was a different project, and one with decidedly limited possibilities.

Another notion that is up for grabs, besides nominalism, is that of ontological commitment itself. You speak of reducing ontology by extending logic. A trivial example in the literature is the assuming of a primitive ancestral functor to obviate the use of classes in defining the ancestral. This is why, in fixing one standard of ontological commitment, I have pinned the standard to first-order predicate logic. If modal operators or substitutional quantification or a primitive ancestral are assumed, then the ontology in my sense becomes indeterminate except relative to one or another chosen translation, if such there be, into first-order logic and interpreted predicates.

Professor Hartry Field   Page 2

    Thus an extending of logic is not a reducing of ontology, in my sense of ontology, but only a changing of the subject: a turning of one's back on my ontological question, pending a choice of translation back into standard form. But the fact remains that ontology in this sense is but one of a potential family of related concepts, no less deserving of the same name with distinctive qualifiers.

Sincerely yours,

W. V. Quine

WVQ:jt

Preface to First Edition

This monograph represents what I believe to be a new approach to the philosophy of mathematics. Most of the literature in the philosophy of mathematics takes the following three questions as central:

(a) How much of standard mathematics is true? For example, are conclusions arrived at using impredicative set theory true?
(b) What entities do we have to postulate to account for the truth of (this part of) mathematics?
(c) What sort of account can we give of our knowledge of these truths?

A fourth question is also sometimes discussed, though usually quite cursorily:

(d) What sort of account is possible of how mathematics is applied to the physical world?

Now, my view is that question (d) is the really fundamental one. And by focussing on the question of application, I was led to a surprising result: that to explain even very complex applications of mathematics to the physical world (for instance, the use of differential equations in the axiomatization of physics) it is not necessary to assume that the mathematics that is applied is true, it is necessary to assume little more than that mathematics is consistent. This conclusion is not based on any general instrumentalist stratagem: rather, it is based on a very special feature of mathematics that other disciplines do not share.

The fact that the application of mathematics doesn't require that the mathematics that is applied be true has important implications for the philosophy of mathematics. For what good argument is there for regarding standard mathematics as a body of truths? The fact that standard mathematics is logically derived from an apparently consistent body of axioms isn't enough; the question is, why regard the axioms as *truths*, rather than as fictions that for a variety of reasons mathematicians have become interested in? The only non-question-begging arguments I have ever heard for the view that mathematics is a body of truths all rest ultimately on the applicability of mathematics to the physical world; so if applicability to the physical world isn't a good argument either, then there is no reason to regard any part of mathematics as true. This is not of course to say that there is something wrong with mathematics; it's simply to say that mathematics isn't the sort of thing that can be appropriately evaluated in terms of truth and falsehood. Questions (a)–(c) are thus trivially answered: no part of mathematics is true (but you can use impredicative reasoning and other controversial reasoning all you like in mathematics as long as you're pretty sure it's consistent); consequently no entities have to be postulated to account for mathematical truth, and the problem of accounting for the knowledge of mathematical truths vanishes. (Of course, the problem of accounting for our knowledge of what mathematical conclusions follow from what mathematical premises still remains. But that is *logical* knowledge, not *mathematical* knowledge: it isn't knowledge of any special realm of mathematical entities.)*

The hardest part of showing that the application of mathematics doesn't require that the mathematics that is applied be true is to show that mathematical entities are theoretically dispensable in a way that theoretical entities in science are not: that is, that one can always re-axiomatize scientific theories so that there is no reference to or quantification over mathematical entities in the reaxiomatization (and one can do this in such a way that the resulting axiomatization is fairly simple and attractive). To show convincingly that such nominalistic reaxiomatizations of serious physical theories are possible requires a rather detailed technical argument. In this monograph I have in fact given such an argument (in the case of one physical theory I judge to be fairly typical). But I have tried to make the main ideas of my approach accessible to those without the background or the patience to follow all of the technical details.

The motivation for this project did not come solely from considerations about the philosophy of mathematics or about ontology: certain ideas in the philosophy of science (such as the desirability of what I call 'intrinsic explanations' and the desirability of eliminating certain sorts

---

* In these first two paragraphs I have used the term 'mathematics' a bit more narrowly than in the text: in these paragraphs, only sentences containing terms referring to mathematical entities or variables ranging over mathematical entities count as part of mathematics. (Compare note 1 of the text.)

of 'arbitrariness' or 'conventional choice' from our ultimate formulation of theories) also played a key role. These ideas from the philosophy of science are touched on in Chapter 5; they yield support, independent of ontological considerations, for the account of the application of mathematics being suggested here. I also discuss (mostly in Chapter 9 but to some extent also in Chapter 4) some issues about logic and about ontological commitment: in particular, the relativity of ontological commitment to the underlying logic, i.e. the fact that one can often reduce one's ontological commitments by expanding one's logic. This is a fact about ontological commitment that has not been sufficiently discussed by philosophers writing on ontological questions, and one of the issues I address myself to in the final chapter is under what circumstances if any it is reasonable to expand one's logic in order to reduce one's ontology.

I would like to thank the University of Southern California, the National Science Foundation and the Guggenheim Foundation for their generous support that provided me with the time needed for research and writing of this project. At a less material level, I would like to thank John Burgess and especially Scott Weinstein for helping me to get straight the relation between the consistency of mathematics and its conservativeness (cf. the Appendix to Chapter 1); and to Burgess, Tony Martin, and Yiannis Moschovakis for helpfully answering various questions that arose when I attempted to prove a false claim about the system N_0 that is discussed in Chapter 9. Several readers of an earlier draft made helpful comments that enabled me to clarify and improve my argument: among them I would like especially to mention Solomon Feferman, Michael Friedman, David Hills, Janet Levin, Colin McGinn, and Charles Parsons. Finally, I would like to express a general indebtedness to Hilary Putnam: in philosophy of mathematics as in much else, his work has deeply influenced the way I think about things, even where (as here) the conclusions we have reached are very different.

Here is a chapter-by-chapter description of what follows:

- **Preliminary Remarks.** (a) States the doctrine to be advocated (and to be called 'nominalism'), namely the view that there are no mathematical entities; (b) sketches the most serious objection that has been made to this doctrine: roughly, that mathematical entities are indispensable to practical affairs and to science; (c) describes the strategy most nominalists have adopted for trying to get around this objection; and (d) describes an alternative strategy for overcoming the objection, which is the strategy to be employed in this book.

- **1 Why the Utility of Mathematical Entities is Unlike the Utility of Theoretical Entities.** In this chapter I argue that it is legitimate to use mathematics to draw nominalistic conclusions (i.e. conclusions statable without reference to mathematical entities) from nominalistic premises, without assuming that the mathematics used in this way is true, but assuming little more than that it is consistent. More precisely, what one assumes about mathematics (and the relationship of this assumption to the assumption that mathematics is consistent is discussed in the Appendix to the chapter) is that mathematics is *conservative*: any inference from nominalistic premises to a nominalistic conclusion that can be made with the help of mathematics could be made (usually more long-windedly) without it. This is a fundamental difference between the use of mathematical entities and the use of the theoretical entities of science: no such conservativeness property holds for the latter.

  The utility of theoretical entities in science is due solely to their *theoretical indispensability*: without theoretical entities, no (sufficiently attractive) theory is possible. At first blush, it appears that mathematical entities are theoretically indispensable too, for they seem to be needed in axiomatizing science; it appears, then, that the conservativeness of mathematics accounts for only part of its utility. In later chapters, however, I argue that mathematical entities are not theoretically indispensable, and that the entire utility of mathematics can be accounted for by its conservativeness, without assuming its truth.

- **2 First Illustration of Why Mathematical Entities are Useful: Arithmetic.** This chapter and the next provide elementary illustrations of the kind of application of mathematics that can be accounted for by the conservativeness of mathematics alone, without invoking the assumption that the mathematics being applied is true. This chapter concerns the application of the arithmetic of natural numbers.

- **3 Second Illustration of Why Mathematical Entities are Useful: Geometry and Distance.** Here I show that the use of real numbers in geometry can be accounted for by the conservativeness of mathematics, without assuming the truth of the theory of real numbers. This illustration of the ideas of Chapter 1 will play a major role in ensuing chapters. To give a bit more detail: I discuss Hilbert's axiomatization of Euclidean geometry, which, since it doesn't involve real numbers, shows that real numbers are theoretically dispensable in geometry; then I discuss two theorems that Hilbert proved about his axiomatization of geometry, namely his representation and uniqueness theorems, and show how the representation theorem explains the utility of real numbers in geometric reasoning (without requiring that the theory of real numbers be true) while the uniqueness theorem establishes that the axiomatization without numbers has certain quite desirable properties.

- **4 Nominalism and the Structure of Physical Space.** Here it is argued that the Hilbert theory of the previous chapter not only dispenses with real numbers, but is (or can be made with a little rewriting) a genuinely nominalistic theory of the structure of physical space. Arguing this involves a brief discussion of some questions in the philosophy of space and time, and an issue in the philosophy of logic that arises again in Chapter 9.
- **5 My Strategy for Nominalizing Physics, and its Advantages.** Here I suggest that the Hilbert theory of geometry, and its representation and uniqueness theorems, provide a general model of how physical theories are to be nominalized. Several features of Hilbert's version of geometry are cited; it is argued that these features are highly advantageous ones and a decision is made to require of an adequate nominalization of physics that it have analogous advantages. It is also pointed out that the other nominalistic approaches which were contrasted to my approach in the Preliminary Remarks do not lead to physical theories with these advantageous features.
- **6 A Nominalistic Treatment of Newtonian Space-Time.** This chapter extends the Hilbert treatment of space to space-time, emphasizing the advantages of the resulting theory over more usual approaches to space-time. (The key advantages of my approach, aside from its being nominalistic, are that it is more thoroughly 'intrinsic' and (closely related) that it avoids use of a certain kind of 'arbitrary choice' of scale, rest frame, coordinate system, etc.) This is the first of the chapters that have a fairly technical subject matter, but it is written in an informal enough way so that most readers should be able to get the main idea of the approach I am following and its advantages.

- **7 A Nominalistic Treatment of Quantities, and a Preview of a Nominalistic Treatment of the Laws involving them.** Here I discuss very briefly how quantities like temperature are to be dealt with nominalistically. I also outline the strategy that is to be used in the next chapter for dealing nominalistically with laws involving these quantities, such as differential equations. This chapter, like the last, deals with technical material, but is informal enough so that most readers should get the general idea.
- **8 Newtonian Gravitational Theory Nominalized.** This chapter is quite technical: it is a detailed sketch of how one particular theory is to be formulated nominalistically, and how the adequacy of this formulation is to be proved. I suspect that many readers will not be interested in going through the details, but I recommend that they read at least section A: this gives a relatively simple illustration of the same strategy of nominalization that is used in more complicated contexts later on in the chapter.
- **9 Logic and Ontology.** There are two respects in which the treatment of physics in the foregoing chapters goes beyond first-order logic, and this final chapter discusses what morals are to be drawn from this. It is argued first that this extra logic does not violate nominalism; second, that use of this extra logic is preferable to use of set theoretic surrogates for the logic (which *would* violate nominalism); third, that use of this extra logic is probably dispensable anyway. The first two of these points involve issues about ontological commitment that are of interest independently of the theory being presented in this monograph.

Preliminary Remarks

Nominalism is the doctrine that there are no abstract entities. The term 'abstract entity' may not be entirely clear, but one thing that does seem clear is that such alleged entities as numbers, functions, and sets are abstract—that is, they would be abstract if they existed. In defending nominalism, therefore, I am denying that numbers, functions, sets, or any similar entities exist.

Since I deny that numbers, functions, sets, etc. exist, I deny that it is legitimate to use terms that purport to refer to such entities, or variables that purport to range over such entities, in our ultimate account of what the world is really like.

This appears to raise a problem: for our ultimate account of what the world is really like must surely include a physical theory; and in developing physical theories one needs to use mathematics; and mathematics is full of such references to and quantifications over numbers, functions, sets, and the like. It would appear then that nominalism is not a position that can reasonably be maintained.

There are a number of *prima facie* possible ways to try to resolve this problem. The way that has proved most popular among nominalistically inclined philosophers is to try to *reinterpret* mathematics—reinterpret it so that its terms and quantifiers don't make reference to abstract entities (numbers, functions, etc.) but only to entities of other sorts, say physical objects, or linguistic expressions, or mental constructions.

My approach is different: I do not propose to reinterpret any part of classical mathematics; instead, I propose to show that the mathematics needed for application to the physical world does not include anything which even *prima facie* contains references to (or quantifications over) abstract entities like numbers, functions, or sets. Towards that part of mathematics which does contain references to (or quantifications over) abstract entities—and this includes virtually all of conventional mathematics—I adopt a fictionalist attitude: that is, I see no reason to regard this part of mathematics as *true*.[1]

Most recent philosophers have been hostile to fictionalist interpretations of mathematics, and for good reason. If one *just* advocates fictionalism about a portion of mathematics, without showing how that part of mathematics is dispensable in applications, then one is engaging in intellectual doublethink: one is merely taking back in one's philosophical moments what one asserts in doing science, without proposing an alternative formulation of science that accords with one's philosophy. This (Quinean) objection to fictionalism about mathematics can only be undercut by showing that there is an alternative formulation of science that does not require the use of any part of mathematics that refers to or quantifies over abstract entities. I believe that such a formulation is possible; consequently, without intellectual doublethink, I can deny that there are abstract entities.

The task of showing that one can reformulate all of science so that it does not refer to or quantify over abstract entities is obviously a very large one; my aim in this monograph is only to illustrate what I believe to be a new strategy toward realizing this goal, and to make both the goal and the strategy look attractive and promising. My attempt to make the strategy look promising ultimately takes the following form: I show, in Chapter 8, how in the context of certain physical theories (field theories in flat space-time[2]) one can develop an analogue of the calculus of several real variables that does not quantify over real numbers or functions or any such thing. Although I do not develop this analogue of calculus completely (e.g. I do not discuss integration), I do sketch enough of it to show how a nominalistic version of the Newtonian theory of gravitation could be given. This nominalistic version of gravitational theory has all the nominalistically statable consequences of the usual platonistic (i.e. non-nominalistical) versions of the theory. Moreover, I believe that the nominalistic reformulation is mathematically attractive, and that there

---

[1] The "part of mathematics that doesn't contain references to abstract entities" is really just applied logic: it is the systematic deduction of consequences from axiom systems (axiom systems similar in many respects to those used in platonistic mathematics, but containing references only to physical entities). Very little of ordinary mathematics consists merely of the systematic deduction of consequences from such axiom systems: my claim, however, is that ordinary mathematics can be replaced in application by a new mathematics which does consist only of this.

[2] I believe the approach is generalizable to curved space-time, but haven't thought through all the details.

are considerations other than ontological ones that favour it over the usual platonistic formulations.

I must admit that the formulation of gravitational theory which I arrive at will not satisfy every nominalist: I use several devices which some nominalists would question. In particular, nominalists with any finitist or operationalist tendencies will not like the way I formulate physical theories, for my formulations will be no more finitist or operationalist than the usual platonistic formulations of these theories are. To illustrate the distinction I have in mind between nominalist concerns on the one hand and finitist or operationalist concerns on the other, consider an example. Someone might object to asserting that between any two points of a light ray (or an electron, if electrons have non-zero diameter) there is a third point, on the ground that this commits one to infinitely many points on the light ray (or the electron), or on the ground that it is not in any very direct sense checkable. But these grounds for objecting to the assertion are not nominalistic grounds as I am using the term 'nominalist', for they arise not from the nature of the postulated entities (viz. the parts of the light ray or of the electron) but from the structural assumptions involving them (viz. that there are infinitely many of them in a finite stretch). I am not very impressed with finitist or operationalist worries, and consequently I make no apologies for making some fairly strong structural assumptions about the basic entities of gravitational physics in what follows. It is not that I have no sympathy whatever for the program of reducing the structural assumptions made about the entities postulated in physical theories—if this can be done, it is interesting. But as far as I aware, it has not been successfully done even in platonistic formulations of physics: that is, no platonistic physics is available which uses a mathematical system less rich than the real numbers to represent the positions of the parts of a light ray or of an electron. Consequently, although I will make it a point not to make any structural assumptions about entities beyond the structural assumptions made in the usual platonistic theories about these entities, I will also feel no compulsion to reduce my structural assumptions below the platonistic level.[3] The reduction of structural assumptions is simply not my concern.

---

[3] As it happens, a certain reduction of structural assumptions will fall out 'by accident', on one of the two nominalistic formulations of gravitational theory I will give (the one I will call N_0 in Chapter 9). Moreover, both nominalistic formulations, but especially N_0, seem especially well suited for a study of the effects of further weakenings of the structural assumptions.

Although I feel no apologies are in order for my use of structural assumptions that would offend the finitist or operationalist, there is another device I have used which I do feel slightly apologetic about. But I try to argue in the final chapter that it is less objectionable than it might at first seem, and that it is probably eliminable anyway.

I would like to make clear at the outset that nothing in this monograph purports to be a positive argument for nominalism. My goal rather is to try to counter the most compelling arguments that have been offered against the nominalist position. It seems to me that the only non-question-begging arguments against the sort of nominalism sketched here (that is, the only non-question-begging arguments *for* the view that mathematics consists of *truths*) are all based on the applicability of mathematics to the physical world. Notice that I do not say that the only way to argue that *a given mathematical axiom* is true is on the basis of *its* application to the physical world: that would be incorrect. For instance, if one grants that the elementary axioms of set theory are true, one can with at least some plausibility argue for the truth of the axiom of inaccessible cardinals on the grounds that this axiom accords with the general conception of sets that underlies the more elementary axioms. More generally, if we assume that the concept of truth has non-trivial application in at least one part of pure mathematics (or to be more precise, if we assume that there is at least one body of pure mathematical assertions that includes existential claims and that is true), then we are assuming that there are mathematical entities. From this we can conclude that there must be some body of facts about these entities, and that not all facts about these entities are likely to be relevant to known applications to the physical world; it is then plausible to argue that considerations other than application to the physical world, for example, considerations of simplicity and coherence within mathematics, are grounds for accepting some proposed mathematical axioms as true and rejecting others as false. This is all fine; but it is of relevance only after one grants the assumption that for some part of mathematics the concept of truth has non-trivial application, and this is an assumption that the nominalist will not grant.

There can be no doubt that the axioms of, say, real numbers are important, and that they are non-arbitrary; and an explanation of their non-arbitrariness, based on their applicability to the physical world but compatible with nominalism, will be given in Chapters 1–3. The present point is simply that from the importance and non-arbitrariness of these axioms, it doesn't obviously follow that these axioms are true, i.e. it doesn't obviously follow that there are mathematical entities that these axioms correctly describe. The existence of such entities may in the end be a reasonable conclusion to draw from the importance and non-arbitrariness of the axioms, but this needs an argument. When the debate is pushed to this level, I believe it becomes clear that there is one and only one serious argument for the existence of mathematical entities, and that is the Quinean argument that we need to postulate such entities in order to carry out ordinary inferences about the physical world and in order to do science.[4] Consequently it seems to me that if I can undercut this argument for the existence of mathematical entities, then the position that there are such entities will look like unjustifiable dogma.

The fact that what I am trying to do is not to provide a positive argument for nominalism but to undercut the only available argument for platonism must be borne in mind in considering an important methodological issue. Although in this monograph I will be espousing nominalism, I am going to be using platonistic methods of argument: I will for instance be proving *platonistically*, not nominalistically, that a certain nominalistic theory of gravitation has all of the nominalistically-statable consequences that the usual platonistic formulation of the Newtonian theory of gravitation has. It might be thought that there is something wrong about using platonistic methods of proof in an argument for nominalism. But there is really little difficulty here: if I am successful in proving *platonistically* that abstract entities are not needed for ordinary inferences about the physical world or for science, then anyone who wants to argue for platonism will be unable to rely on the Quinean argument that the existence of abstract entities is an indispensable assumption. The monograph shows that any such argument would be inconsistent with the platonistic position that is being argued for. The would-be platonist, then, will be forced into either accepting abstract objects without argument or else relying on other arguments for platonism, arguments which in my opinion are quite unpersuasive. The upshot

---

[4] The most thorough presentation of the Quinean argument is actually not by Quine but by Hilary Putnam (Putnam 1971, especially chs. 5–8). Some of the arguments I do not take seriously (e.g. the argument that we need to postulate mathematical entities in order to account for mathematical intuitions) are well treated in ch. 2 of Chihara 1973.

then (if I am right in my negative appraisal of alternative arguments for platonism) is that platonism is left in an unstable position: it entails its own unjustifiability.[5]

It may be of course that my negative appraisal of alternative arguments for platonism is wrong. Interestingly enough, the platonist who bases his case for platonism on some such alternative argument may even find what I have to say welcome; for independently of nominalistic considerations, I believe that what I do here gives an attractive account of how mathematics is applied to the physical world. This is I think in sharp contrast to many other nominalistic doctrines, e.g. doctrines which reinterpret mathematical statements as statements about linguistic entities or about mental constructions. Such nominalistic doctrines do nothing toward illuminating the way in which mathematics is applied to the physical world. (I will return to this point in Chapter 5.)

---

[5] Actually, I do not think that a platonistic proof of the adequacy of our theories serves *merely* as a *reductio*: I think that a nominalist too should be convinced by a platonistic proof about the deductive powers of a given nominalistic theory. But a defense of this claim would be a long story. (Some much too brief remarks on this matter are contained in note 10 in the next chapter.) In any case, the nominalist need not ultimately rely on such platonistic proofs of the adequacy of his systems: in principle at least, he and his fellow nominalists could simply spin out deductions from nominalistic axiom systems like the ones suggested later in the monograph. In this sense, the reliance on platonistic proofs could be regarded as a temporary expedient.

1
Why the Utility of Mathematical Entities is Unlike the Utility of Theoretical Entities

No one can sensibly deny that the invocation of mathematical entities in some contexts is useful. The question arises as to whether the utility of mathematical existence-assertions gives us any grounds for believing that such existence-assertions are true. I claim that in answering this question one has to distinguish two different ways in which mathematical existence-assertions might be useful; I grant that if such assertions are useful in a certain respect, then that would indeed be evidence that they are true; but the most obvious respect in which mathematical existence assertions are useful is, I claim, quite a different one, and I will argue that the utility of such assertions in this respect gives no grounds whatever for believing the assertions to be true.

To be more explicit, I will argue that the utility of mathematical entities is structurally disanalogous to the utility of theoretical entities in physics. The utility of theoretical entities lies in two facts:

(a) they play a role in powerful theories from which we can deduce a wide range of phenomena; and
(b) no alternative theories are known or seem at all likely which explain these phenomena without similar entities.

[The unsympathetic reader may dispute (b): if any body of sentences counts as a 'theory' and any deduction from such a 'theory' counts as an explanation, then there clearly are alternatives to the usual theories of subatomic particles: e.g., take as your 'theory' the set T* of all the consequences of T that don't contain reference to subatomic particles (where T is one of the usual theories that does contain reference to subatomic particles); or if you want a recursively axiomatized 'theory', let T** be the Craigian reaxiomatization of the theory T* just described. Since I don't know any formal conditions to impose which would rule out such bizarre trickery, let me simply say that by 'theory' I mean *reasonably attractive theory*; 'theories' like T* and T** are obviously uninteresting, since they do nothing whatever toward explaining the phenomena in question in terms of a small number of basic principles.]

The upshot of (a) and (b) is that subatomic particles are *theoretically indispensable*; and I believe that that is as good an argument for their existence as we need. Now, later on in the monograph I will argue that mathematical entities are not theoretically indispensable: although they do play a role in the powerful theories of modern physics, we can give attractive reformulations of such theories in which mathematical entities play no role. If this is right, then we can safely adhere to a fictionalist view of mathematics, for adhering to such a view will not involve depriving ourselves of a theory that explains physical phenomena and which we can regard as literally true.

But the task of arguing for the theoretical dispensability of mathematical entities is a matter for later. What I want to do now is to give an account, *consistent with* the theoretical dispensability of mathematical entities, of why it is useful to make mathematical existence-assertions in certain contexts.

The explanation of why mathematical entities are useful involves a feature of mathematics that is not shared by physical theories that postulate unobservables. To put it a bit vaguely for the moment: if you take any body of nominalistically stated assertions N, and supplement it with a mathematical theory S, you don't get any nominalistically stated conclusions that you wouldn't get from N alone. The analog for theories postulating subatomic particles is of course not true: if T is a theory that involves subatomic particles and is at all interesting, then there are going to be lots of cases of bodies P of wholly macroscopic assertions which in conjunction with T yield macroscopic conclusions that they don't yield in absence of T; if this were not so, theories about subatomic particles could never be tested.

I'll state these claims more precisely in a moment, but first I should say that the claim about mathematics would be almost totally trivial if mathematics consisted only of theories like number theory or pure set theory, i.e. set theory in which no allowance is made for sets with members that are not themselves sets. But these theories are by themselves of no interest from the point of view of applied mathematics, for there is no way to apply them to the physical world. That is, there is no way in which they are even *prima facie* helpful in enabling us to deduce nominalistically statable consequences from nominalistically statable premises. In order to be able to apply any postulated abstract entities to the physical world, we need *impure* abstract entities, e.g. functions that map physical objects into pure abstract entities. Such impure abstract entities serve as a bridge between the pure abstract entities and the physical objects; without the bridge, the pure objects would be idle. Consequently, if we regard functions as sets of a certain sort, then the mathematical theories we should be considering must include at least a minimal amount of impure set theory: set theory that allows for the possibility of urelements, where an urelement is a non-set which can be a member of sets. In fact, in order to be sufficiently powerful for most purposes, the impure set theory must differ from pure set theory not only in allowing for the possibility of urelements, it must also allow for non-mathematical vocabulary to appear in the comprehension axioms (i.e. in the instances of the axiom schema of separation or of replacement). So the 'bridge laws' must include laws that involve the mathematical vocabulary and the physical vocabulary together.

Something rather analogous is true of the theory of subatomic particles. One can artificially formulate such a theory so that none of the non-logical[6] vocabulary that is applied to observable physical objects is applied to the subatomic particles; in general it seems to me pointless to formulate physical theories in this way, but to press the analogy with the mathematical case as far as it will go, let us suppose it done. If it is done, and if we suppose that T is a physical theory stated entirely in this vocabulary, then of course, it *will* be the case that if we add T to a bunch of macroscopic assertions P, we will be able to derive no results about observables that weren't derivable already. But that is for a wholly uninteresting reason: it is because the theory T by itself is not even *prima facie* helpful in deducing claims about observables from other claims about observables. In order to make it even *prima facie* helpful, we have to add 'bridge laws', laws which connect up the entities

---

[6] Count '=' as logical.

and/or the vocabulary of the (artificially formulated) physical theory with observables and the properties by which we describe them. So far, then, like the mathematical case.

But there is a fundamental difference between the two cases, and that difference lies in the nature of the bridge laws. In the case of subatomic particles, the theory T, interpreted now so as to include the bridge laws (and perhaps also some assumptions about initial conditions), can be applied to bodies of premises about observables in such a way as to yield genuinely new claims about observables, claims that would not be derivable without T. Whereas in the mathematical case the situation is very different. Here, if we take a mathematical theory that includes bridge laws (i.e. includes assertions of the existence of functions from physical objects into 'pure' abstract objects, including perhaps assertions obtained via a comprehension principle that uses mathematical and physical vocabulary in the same breath), then that mathematics is applicable to the world, i.e. it is useful in enabling us to draw nominalistically statable conclusions from nominalistically statable premises. *But here, unlike in the case of physics, the conclusions we arrive at by these means are not genuinely new, they are already derivable in a more long-winded fashion from the premises, without recourse to the mathematical entities.*

This claim, unlike the one I will make later about the theoretical dispensability of mathematical entities, is pretty much of an incontrovertible fact, but one very much worth emphasizing. So first let me state the point more precisely than I have done.

A first stab at putting the point precisely would be to say that for any mathematical theory S and any body of nominalistic assertions N, N + S is a conservative extension of N. However, this formulation isn't quite right, and it is worth taking the trouble to put the point accurately. The problem with this formulation is that since N is a nominalistic theory, it may say things that *rule out* the existence of abstract entities, and so N + S may well be inconsistent. But it is clear how to deal with this: first, introduce a 1-place predicate '$M(x)$', meaning intuitively 'x is a mathematical entity'; second, for any nominalistically stated assertion A, let A^* be the assertion that results by restricting each quantifier of A with the formula 'not $M(x_i)$' (for the appropriate variable 'x_i');[7] and third,

---

[7] That is, replace every quantification of form '$\forall x_i(\ldots)$' by '$\forall x_i(\text{if not } M(x_i) \text{ then}\ldots)$', and every quantification of form '$\exists x_i(\ldots)$' by '$\exists x_i(\text{not } M(x_i) \text{ and}\ldots)$'.

for any nominalistically stated body of assertions N, let N* consist of all assertions A^* for A in N. N* is then an 'agnostic' version of N: for instance, if N says that all objects obey Newton's laws, then N* says that all *non-mathematical* objects obey Newton's laws, but it allows for the possibility that there are mathematical objects that don't. (Actually N* is in one respect too agnostic: in ordinary logic we assume for convenience that there is at least one thing in the universe, and in the context of a theory like N this means that there is at least one non-mathematical thing. So it is really N*+ '$\exists x \neg M(x)$' that gives the agnostic content of N.)

Whether a similar point needs to be made for our mathematical theory S depends on what we take S to be. If S is simply set theory allowing for urelements, no restriction on the variables is needed, since the theory already purports to be about non-sets as well as sets: we merely need to connect up the notion of set that occurs in it with our predicate 'M', by adding the axiom '$\forall x(Set(x) \to M(x))$'. If in addition the mathematical theory includes portions like number theory, considered as independent disciplines unreduced to set theory, then we must restrict all variables in them by a new predicate 'Number', and add the axioms '$\forall x(Number(x) \to M(x))$' and '$\exists x(Number(x))$'. Presumably, however, everyone agrees that mathematical theories really ought to be written in this way (that is, presumably no one believes that all entities are mathematical), so I will not introduce a special notation for the modified version of S, I'll assume that S is written in this form from the start. (The analogous assumption for N would be inappropriate: the nominalist wants to assert not N*, but the stronger claim N.)

Having dealt with these tedious points, I can now state accurately the claim made at the end of the next to last paragraph.

Principle C (for 'conservative'): Let A be any nominalistically stable[8] assertion, and N any body of such assertions; and

---

[8] The formal content of saying that N is 'nominalistically stable' is simply that it not employ the special mathematical vocabulary of the mathematical theory to be introduced, including 'mathematical'. This is all we need to build into 'nominalistically stable' in order for Principle C to be true. For Principle C to be *of interest*, we must suppose in addition that the intended ontology of N does not include any entities in the intended extension of the predicate 'M' of S; for if this condition were violated, then N* + S would not correspond to the 'intended' way of combining N and S.

[*This footnote is changed from the 1st edition.* There, instead of the first sentence above, I incorrectly wrote: "The formal content of saying that N is 'nominalistically stable' is

let S be any mathematical theory. Then A^* isn't a consequence of $N^* + S +$ '$\exists x \neg M(x)$' unless A is a consequence of N.

Why should we believe this principle? Well, it follows[9] from a slightly stronger principle that is perhaps a bit more obvious:

Principle C': Let A be any nominalistically statable assertion, and N any body of such assertions. Then A^* isn't a consequence of $N^* + S$ unless it is a consequence of N^* alone.

This in turn is equivalent (assuming the underlying logic to be compact) to something still more obvious-sounding:

Principle C'': Let A be any nominalistically statable assertion. Then A^* isn't a consequence of S unless it is logically true.

Now I take it to be perfectly obvious that our mathematical theories do satisfy Principle C''. After all, these theories are commonly regarded as being 'true in all possible worlds' and as '*a priori* true'; and though these characterizations of mathematics may be contested, it is hard to see how any knowledgeable person could regard our mathematical theories in these ways if those theories implied results about concrete entities alone that were not logically true. The same argument can be used to directly motivate Principle C', thereby obviating the need of the compactness assumption: if mathematics together with a body N^* of nominalistic assertions implied an assertion A^* which wasn't a logical consequence of N^* alone, then the truth of the mathematical theory would hinge on the logically consistent body of assertions $N^* + \neg A^*$ not being true. But it would seem that it must be possible, and/or not *a priori* false, that such a consistent body of assertions about concrete objects alone is true; if so,

---

simply that it not overlap in non-logical vocabulary with the mathematical theory to be introduced. (Recall that '=' counts as logical.)" This obviously conflicts with the remarks above on the need of impure mathematics. For discussion, see the end of subsection 4.1 of the new Preface.]

[9] Proof: Suppose $N^* + S +$ '$\exists x \neg M(x)$' implies A^*. Then $N^* + S$ implies $A^* \vee \forall x(\neg M(x) \rightarrow x \neq x)$; that is, it implies B^* where B is $A \vee \forall x(x \neq x)$. Applying Principle C', we get that N^* implies B^*, and consequently that $N^* +$ '$\exists x \neg M(x)$' implies A^*. From this it clearly follows that N implies A.

Principle C' does not quite follow from Principle C, for a theory S could imply that there are non-mathematical objects but not imply anything else about the non-mathematical realm (in particular, not imply that there are at least two mathematical objects—the latter would violate Principle C as well as Principle C').

then the failure of Principle C would show that mathematics couldn't be 'true in all possible worlds' and/or '*a priori* true'. The fact that so many people think it does have these characteristics seems like some evidence that it does indeed satisfy Principle C′ and therefore Principle C.

This argument isn't conclusive: standard mathematics might turn out not to be conservative (i.e. not to satisfy Principle C), for it *might* conceivably turn out to be inconsistent, and if it is inconsistent it certainly isn't conservative. We would however regard a proof that standard mathematics was inconsistent as extremely surprising, and as showing that standard mathematics needed revision. Equally, it would be extremely surprising if it were to be discovered that standard mathematics implied that there are at least 10^{60} non-mathematical objects in the universe, or that the Paris Commune was defeated; and were such a discovery to be made, all but the most unregenerate rationalist would take this as showing that standard mathematics needed revision. *Good* mathematics *is* conservative; a discovery that accepted mathematics isn't conservative would be a discovery that it isn't good.

Indeed, as some of the mathematical arguments in the Appendix to this chapter show, the gap between the claim of consistency and the full claim of conservativeness is, in the case of mathematics, a very tiny one. In fact, for pure set theory, or for set theory that allows for impure sets but doesn't allow empirical vocabulary to appear in the comprehension axioms, the conservativeness of the theory follows from its consistency alone. For full set theory this is not quite true; but a large part of the content of the conservativeness claim for full set theory (probably the only part of the content that is important in application) follows from the consistency of set theory alone (and still more of the content follows from slightly stronger assumptions, like that full set theory is ω-consistent). These claims are demonstrated in the Appendix to this chapter. In any case, I think that the two previous paragraphs show that the same sort of quasi-inductive grounds we have for believing in the consistency of mathematics extend to its conservativeness as well. As we saw earlier, this means that there is a marked disanalogy between mathematical theories and physical theories about unobservable entities: physical theories about unobservables are certainly not conservative, they give rise to genuinely new conclusions about observables.

What the facts about mathematics I have been emphasizing here show is that even someone who doesn't believe in mathematical entities is free to use mathematical existence-assertions in a certain limited context: he can use them freely in deducing nominalistically stated consequences from nominalistically stated premises. And he can do this not because he thinks those intervening premises are true, but because he knows that they preserve truth among nominalistically stated claims.[10]

---

[10] In what sense does he know this? At the very least, he knows it in the sense that a platonist mathematician who proves a result in recursive function theory by means of Church's thesis knows that he could construct a proof that didn't invoke Church's thesis. The platonist mathematician hasn't proved using the basic forms of argument that he accepts that such a proof is possible, for he hasn't proved Church's thesis. (Nor can he even state Church's thesis except by vague terms like 'intuitively computable'.) Still, there is a perfectly good sense in which our platonist mathematician does know that a proof without Church's thesis is possible—after all, he could probably come up with Turing machine programs at each point where Church's thesis was invoked, if given sufficient incentive to do so. In precisely the same sense, the nominalist knows that for any platonist proof of a nominalistically stated conclusion from nominalistically stated premises there is a nominalistic proof of the same thing.

Just what this sense of 'know' is (or, just what *kind* of knowledge is involved) is a difficult matter: it doesn't seem to me quite right to call it 'inductive' knowledge. But however this may be, it is a kind of knowledge (or justification) whose strength can be increased by inductive considerations: in the recursive function case, by knowledge that in the past one had been able to transform proofs involving the imprecise notion of 'intuitively computable' to proofs not involving it when one has tried (or by knowledge that others have been able to effect such transformations, and that one's own judgements of intuitive computability tend to coincide with theirs). In the conservativeness case, the kind of inductive considerations that are relevant are the knowledge that in the past no one has found counterexamples to conservativeness, and also the knowledge that in many actual cases where platonistic devices are used in proofs of nominalistic conclusions from nominalistic premises (such as the cases discussed in Chapters 2 and 3), these devices are eliminable in what seems to be a more or less systematic way.

These remarks suggest that the nominalistic position concerning the use of platonistic proofs is about comparable to the platonist's position concerning proofs that use Church's thesis. Actually I think that the nominalist's position is in one respect even better, for he can rely on something that the platonistic recursion theorist has no analog of: viz., the mathematical arguments for conservativeness given in the Appendix. Of course, these arguments don't raise the claim that mathematics is conservative to complete certainty, for two reasons. One reason is that something at least as strong as the consistency of set theory is assumed in them, and no one (platonist or nominalist) can be *completely* sure of that. The other reason is that these proofs (at least the first, and both if one is sufficiently strict about what counts as nominalist) are platonistic, and so some story has to be told about how the nominalist is justified in appealing to them outside the context of a *reductio*. I think some such story can be told, but it would be a long one. (An essential idea of the story would be that we use conservativeness to argue for conservativeness: we've seen that the nominalist has various initial quasi-inductive arguments which support the conclusion that it is safe to use mathematics in certain contexts; if he then *using mathematics in one of those contexts* can prove that it is safe to use mathematics in those contexts, this can raise the support of the initial conclusion quite substantially.)

This point is not of course intended to license the use of mathematical existence assertions in axiom systems for the particular sciences: such a use of mathematics remains, for the nominalist, illegitimate. (Or more accurately, a nominalist should treat such a use of mathematics as a temporary expedient that we indulge in when we don't know how to axiomatize the science properly, and that we ought to try to eliminate.) The point I am making, however, does have the consequence that *once such a nominalistic axiom system is available*, the nominalist is free to use any mathematics he likes for deducing consequences, as long as the mathematics he uses satisfies Principle C.

So if we ignore for the moment the role of mathematics in axiomatizing the sciences, then it looks as if the satisfaction of Principle C is the really essential property of mathematical theories. The fact that mathematical theories have this property is doubtless one motivation for the platonist's assertion that such theories are 'true in all possible worlds'. It does not appear to me, however, that the satisfaction of Principle C provides reason for regarding a theory as true at all (even in the actual world). Certainly such speculations, typical of extreme platonism, as to for instance whether the continuum hypothesis is 'really true', seem to lose their point once one recognizes conservativeness as the essential requirement of mathematical theories: for the usual Gödel and Cohen relative consistency proofs about set theory plus the continuum hypothesis and set theory plus its denial are easily modified into relative *conservativeness* proofs. In other words, assuming that standard set theory satisfies Principle C, so does standard set theory plus the continuum hypothesis and standard set theory plus its denial; so it follows that *either theory could be used without harm in deducing consequences about concrete entities from nominalistic theories.*

The same point made about the continuum hypothesis holds as well for less *recherché* mathematical assertions. Even standard axioms of number theory can be modified without endangering Principle C; similarly for standard axioms of analysis. What makes the mathematical theories we

---

A platonist might be inclined to dismiss the sort of quasi-inductive knowledge discussed in this note. But to do so would be to pay a high price: *most* of mathematics is known only in this quasi-inductive sort of way. For most of it is proved by rather informal proofs; and though we all do in an important sense *know* that we could reconstruct such proofs formally if forced to do so, still the principle that formal proofs are always possible when we have an intuitively acceptable proof is, like Church's thesis, a principle that we haven't proved and have no prospect of proving.

accept better than these alternatives to them is not that they are true and the modifications not true, but rather that they are more *useful*: they are more of an aid to us in drawing consequences from those nominalistic theories that we are interested in. If the world were different, we would be interested in different nominalistic theories, and in that case some of the alternatives to some of our favorite mathematical theories might be of more use than the theories we now accept.[11] Thus mathematics is in a sense empirical, but only in the rather Pickwickian sense that is an empirical question as to which mathematical theory is useful. It is in an equally Pickwickian sense, however, that mathematical axioms are *a priori*: they are not *a priori* true, for they are not true at all.

The view put forward here has considerable resemblance to the logical positivist view of mathematics. One difference that is probably mostly verbal is that the positivists usually described pure mathematics as analytically true, whereas I have described it as not true at all; this difference is probably mostly verbal, given their construal of 'analytic' as 'lacking factual content'. A much more fundamental difference is that what worried the positivists about mathematics was not so much its postulation of entities as its apparently non-empirical character, and this was a problem not only for mathematics, but for logic as well. Hence they regarded *logic* as analytic or contentless in the same sense that *mathematics* was. I believe that this prevented them from giving any clear explanation of what the 'contentlessness' of mathematics (or of that part of mathematics that quantifies over abstract entities) consists in. The idea of calling a logical or mathematical assertion 'contentless' was supposed to be that a conclusion arrived at by a logical or mathematical argument was in some sense 'implicitly contained in' the premises: in this way, the conclusion of such an argument was 'not genuinely new'. Unfortunately, no clear explanation of the idea that the conclusion was 'implicitly contained in' the premises was ever given, and I do not believe that any clear explanation is possible.

What I have tried to do in this chapter is to show how by giving up (or saving for separate explication) the claim that *logic* (and that part of math that *doesn't* make reference to abstract entities) doesn't yield genuinely new conclusions, we can give a clear and precise sense to the idea that *mathematics* doesn't yield genuinely new conclusions: more precisely, we

---

[11] We will see, however, that the utility of number theory is less subject to such empirical vicissitudes than are theories about say the real numbers.

can show that the part of math that does make reference to mathematical entities can be applied without yielding any genuinely new conclusions about non-mathematical entities.

Appendix: On Conservativeness

It may be illuminating to give two mathematical arguments for the conservativeness of mathematics. The first argument proves, from a set-theoretic perspective (more specifically, from the perspective of ordinary set theory plus the axiom of inaccessible cardinals) that ordinary set theory (and hence standard mathematics, which is reducible to ordinary set theory) is definitely conservative. The second argument is a purely proof-theoretic one: it establishes a slightly restricted form of the conservativeness claim on the basis merely of the assumption that standard set theory is consistent. This is illuminating in showing that the assumption of the conservativeness of set theory is much 'closer to' the assumption that set theory is consistent than to the assumption that it is true.

As a preliminary, let's introduce some notation. Let ZF be standard Zermelo-Fraenkel set theory (including the axiom of choice); let *restricted ZFU* be ZF modified to allow for the existence of urelements, but not allowing for any non-set-theoretic vocabulary to appear in the comprehension axioms (for definiteness, we may stipulate that it contains as an axiom that there is a set of all non-sets). [*Added in this edition:* In accordance with the discussion in the text, we take ZFU to include the assertion that $\forall x(Set(x) \to M(x))$, leaving open whether some urelements (e.g. numbers conceived as non-sets) are also mathematical.] If V is a class of expressions, let ZFU_V be *restricted ZFU* together with any instances of the comprehension schemas in which the vocabulary in V as well as the set-theoretic vocabulary is allowed to appear. What I earlier called 'full set theory' isn't really a single theory: rather, to 'apply full set theory' in the context of a theory T is to apply $ZFU_{V(T)}$, where V(T) is the vocabulary of T. Consequently, what we want to prove is that for any theory T, $ZFU_{V(T)}$ applies conservatively to T. That is, we want to prove

(C_0) If T is any consistent body of assertions not containing 'Math' or '∈' or 'Set', then $ZFU_{V(T)} + T^*$ is also consistent.

[*Changed from 1st edition:* I've added "not containing 'Math' or '∈' or 'Set' "]

18 MATHEMATICAL ENTITIES AND THEORETICAL ENTITIES

(The T here is the N + ¬A of Principle C'.) This in fact will suffice for proving the conservativeness of $ZFU_{V(T)}$ + S, for any standard mathematical theory S: for standard mathematical theories are embeddable in ZF.

So much for preliminaries. How then do we prove that (C_0) holds? The obvious set-theoretic line of proof is this:

> Suppose T is consistent; then it has a model M of accessible cardinality, say with domain D. Pick any entity e not in D. (e is to be thought of as the empty set.) Let D_0 be $D \cup \{e\}$; Let D_1 consist of all non-empty subsets of D_0; let D_2 consist of all non-empty subsets of $D_0 \cup D_1$; and so on. Let D_ω be $D_0 \cup D_1 \cup D_2 \cup \ldots$; let $D_{\omega+1}$ consist of all non-empty subsets of D_ω; and so on. Continuing in this way until you reach an inaccessible cardinal, you get—if certain initial precautions[12] are taken on the choice of D and e—a model of $ZFU_{V(T)}$ + T*. (It is a model of $ZFU_{V(T)}$ + T* rather than merely of ZFU + T* because at each stage you've added *every* set of things available at previous stages.) So $ZFU_{V(T)}$ + T* is consistent.   Q.E.D.

Now let us turn to the proof-theoretic line of argument for conservativeness; the point of doing this is to make clear how narrow the gap is between the consistency of mathematics and its conservativeness.

Indeed, in the case of mathematical theories which don't allow for impure abstract entities (e.g. number theory by itself, or ZF), consistency implies conservativeness: this is an obvious consequence of the Robinson joint consistency theorem.[13] The same result holds also in the more

---

[12] D should either be taken to consist entirely of non-sets, in which case e should be taken to be the empty set (or another non-set); or D should be taken to consist entirely of sets of the same rank and e should be another set of that rank. Given any model of a theory, there is no difficulty in getting another model whose domain meets these conditions.

[13] Suppose S + T* is inconsistent; the Robinson theorem says that there is a sentence B in the language common to S and T* such that $S \vdash B$ and $T^* \vdash \neg B$. Clearly if S and T are both consistent, then B can't be either a logical truth or a contradiction. The language common to S and T* consists, in the case of a 'pure' mathematical theory, of 'M' (the predicate 'mathematical' discussed prior to the formulation of Principle C) and '=', and nothing else. The only statements in this language other than logical truths or contradictions are statements saying how many mathematical objects there are and/or how many non-mathematical objects there are. But since all statements in T* are explicitly restricted to non-mathematical objects, T* can't imply anything about how many mathematical objects there are, and since the mathematical theory is assumed to be a pure one it can't imply anything about how many non-mathematical objects there are. So there can be no such B; that is, the supposition that S and T are consistent but S + T* is inconsistent has been reduced to absurdity.

interesting case of restricted ZFU: here one needs, in addition to the Robinson theorem, the well-known fact that if ZFU is consistent then one can't prove any result about how many non-sets there are.[14] But in the really interesting case of full ZFU, this whole line of argument via the joint consistency theorem is blocked by the fact that the empirical vocabulary that appears in the theory T also appears in set-theoretic comprehension axioms.

The simplest thing to do in this case is to mimic proof-theoretically the set-theoretic argument given three paragraphs back: doing so, it becomes an argument that under certain conditions $ZFU_{V(T)} + T^*$ is interpretable within $ZFU_{V(T)}$, and in fact within ZF. (We don't need the inaccessible cardinal assumption anymore.) If the 'certain conditions' were merely that T is consistent, then we'd know that (C_0) holds as long as ZF is consistent, and this is what we wanted. Unfortunately however we need the stronger assumption that T is provably consistent within ZF; that is, the best we can show is that if ZF is consistent, the following holds:[15]

(C_1) If T is any body of assertions *whose consistency is provable in ZF*, then $ZFU_{V(T)} + T^*$ is consistent.

---

[14] A sketch of the proof of the last fact is given in Jech 1973: 51, problem 1. Using this fact, the proof that conservativeness implies consistency is just as in note 13.

[15] Proof: if ZF is consistent, and ZF ⊢ 'T is consistent' (where 'T is consistent' abbreviates the formalization in ZF of the claim that T is syntactically consistent) then ZF + 'T is consistent' is certainly consistent. Since the Gödel completeness theorem (together with various more elementary facts) is provable in ZF, then so is ZF + 'there is a model of T in which all elements of the domain have the same rank and such that there is a set of that rank that is not in the domain'. (Cf. note 12 for the motivation of this.) If T has n primitive predicates, then a model of T consists of a domain together with n items each corresponding to one of the terms. Introducing new names $b, c_1, \ldots, c_n$ for these things, and a name d for the set of the right rank that isn't in the domain of the model, we see that ZF + '$\langle b, c_1, \ldots, c_n \rangle$ is a model of T' + 'all members of b have the same rank' + 'd has the same rank as all members of b' is also consistent. Call this theory ZF_T.

By the principle of transfinite recursion, there is a formula $\mathfrak{D}(x)$ (in the language of ZF_T) such that

ZF_T (in fact, ZF) ⊢ $\mathfrak{D}(x) \leftrightarrow x \in b \lor x = d \lor (x \neq \emptyset \land \forall y(y \in x \to \mathfrak{D}(y)))$.

If we translate statements of $ZFU_{V(T)} + T^*$ into ZF_T by using $\mathfrak{D}(x)$ to restrict all variables, and translate 'Set(x)' as '$x \notin b$', '∈' as '∈', '∅' as 'd', and '$A_i(x_1, \ldots, x_k)$' where A_i is the ith predicate of T as '$\langle x_1, \ldots, x_k \rangle \in c_i$', then each of the translations of the axioms of $ZFU_{V(T)} + T^*$ is a theorem of ZF_T. Since ZF_T is consistent (on the assumption that ZF is), so is $ZFU_{V(T)} + T^*$.

This is a restricted version of conservativeness: it says that full set theory applies conservatively *to theories which are modellable in ZF*. In actual applications this is probably as much of the conservativeness claim as we ever need. For instance, later on in the book we will want to know that mathematics applies conservatively to a nominalistic version of Newtonian gravitation theory, N_0. But it is completely obvious that if N_0 is consistent then it is modellable in ZF (and the same would presumably be true for other nominalized physical theories); so the conservativeness result we actually need follows merely from the consistency of ZF.

Scott Weinstein (besides clearing up a number of confusions I had gotten into concerning the issues of the last paragraph) pointed out to me that if you strengthen the consistency assumption about ZF slightly, to ω-consistency (or even something a bit weaker than that known as 1-consistency), you can strengthen (C_1) in an attractive way: you can then prove

(C_2) If T is any *consistent and recursively enumerable* body of assertions, then $ZFU_{V(T)} + T^*$ is consistent.[16]

It is all the more obvious that this would be sufficient for practical applications.

Philosophers discussing set theory tend to discuss two of its properties: its consistency, and its (alleged) truth. The argument of this monograph is that the latter is completely irrelevant, and that the former is perhaps a bit too weak—it is too weak unless one is satisfied with (C_1) instead of the full (C_0). [Of course, for the kind of set theory philosophers tend to discuss—*pure* set theory, i.e. ZF—there is no difference at all between consistency and conservativeness (or rather, though they differ conceptually, they are provably equivalent). But pure set theory isn't what is of interest, since

---

[16] To see this, observe first that the preceding note proved a slightly stronger result than was claimed: it proved that if ZF + 'T is consistent' is consistent, then $ZFU_{V(T)} + T^*$ is consistent. So we now need only show that if ZF is ω-consistent and T is consistent and recursively enumerable, then ZF + 'T is consistent' is consistent.

The reason for this is simple: if T is consistent, then nothing is a proof from T of '0 = 1'; and if T is also recursively enumerable, ZF is strong enough to prove '**k** is not the Gödel number of a proof from T of '0 = 1'', for each numeral **k**. By the ω-consistency of ZF it follows that one cannot prove in ZF anything of the form '$\exists x(x$ is the Gödel number of a proof from T of '0 = 1')'; so one can't prove 'T is not consistent' from ZF, and so ZF + 'T is consistent' is consistent.

as remarked before it can never be applied to the physical world, so this is not much of a justification for ignoring conservativeness.] But though we perhaps need to assume a bit more than consistency, we don't need to assume all that much more; and in any case it seems pretty obvious that the stronger property of conservativeness does in fact obtain.

2
First Illustration of Why Mathematical Entities are Useful: Arithmetic

I have explained why it is legitimate for a nominalist to use mathematics in making inferences between nominalistically stated sentences; but I haven't said anything about why or in what way it is *useful* for him to do so. It is important to have a rather vivid understanding of the way that mathematics is useful in such contexts if one is to grasp my strategy for nominalizing physical theories, and so I will devote both this chapter and the next to the matter.

Suppose N is a body of nominalistically stated premises; in the case that will be of primary interest, N will consist of the axioms of a nominalistic formulation of some scientific theory. I think that the key to using a mathematical system S as an aid to drawing conclusions from a nominalistic system N lies in proving in $N^* + S$ the equivalence of a statement in N^* alone with some other statement (which I'll call an *abstract counterpart* of the N^*-statement) which quantifies over abstract entities. Then if we want to determine the validity of an inference in N^* (or equivalently, of an inference in N), it is unnecessary to proceed directly; instead we can if it is convenient 'ascend' from one or more statements in N^* to abstract counterparts of them, then use S to prove from these abstract counterparts an abstract counterpart of some other statement in N^*, and 'descend' back to that statement in N^*. I will illustrate how this procedure works in certain concrete cases; but again I must emphasize that the only thing required for the procedure to be legitimate is not that S be true but merely that $N^* + S$ be a conservative extension of N^*, a condition which will always be met if Principle C of the previous chapter is satisfied.

My first illustration of this general procedure will be a very simple one; here, the mathematical theory S to be applied is simply the arithmetic of natural numbers (or more precisely, arithmetic plus a small amount of set theory, since arithmetic without such things as functions from concrete entities to numbers can never be applied).

Let N be a theory that contains the identity symbol and the usual axioms of identity, but does not contain any terms or quantifiers for abstract objects. In particular, N will not contain singular terms like '87'. It will, however, be convenient to suppose that N contains, besides the usual quantifiers '∀' and '∃', also quantifiers like '$\exists_{87}$' (meaning 'there are exactly 87') and '$\exists_{\geq 87}$' (meaning 'there are at least 87'). The logic is still, of course, recursively axiomatizable—e.g. we could merely add to standard logic the axioms

$$\exists_{\geq 1} x A(x) \leftrightarrow \exists x A(x)$$
$$\exists_{\geq k} x A(x) \leftrightarrow \exists x [A(x) \wedge \exists_{\geq j} y (y \neq x \wedge A(y))],$$

where k is the decimal numeral that immediately succeeds j, and

$$\exists_j x A(x) \leftrightarrow \exists_{\geq j} x A(x) \wedge \neg \exists_{\geq k} x A(x),$$

where k and j are as above. In supposing that N contains this extra structure, we are not enriching either the expressive or the deductive power of N, we are merely ensuring that we can say simply what can be said only in a very roundabout way on the usual but artificial limitation to the two standard quantifiers plus identity. In particular, I must emphasize that by giving N this extra structure, I am not giving it any arithmetic: it contains no singular terms or quantifiers for numbers or any other abstract entities: the numeral '87' occurs in it not as a name, but merely as part of an operator symbol. Our goal is to show how inferences in N can be facilitated by introducing a system S that does contain arithmetic.

To see this, consider the following argument in N:

1. There are exactly twenty-one aardvarks (i.e., $\exists_{21} x A(x)$);
2. On each aardvark there are exactly three bugs;
3. Each bug is on exactly one aardvark; so
4. There are exactly sixty-three bugs.

Is this valid? If one reasons in N, it will take a lot of work to find out—the inference needed for getting from the premises to the conclusion is long and tedious. (Though not nearly as bad as it would have been if we hadn't introduced the numerical quantifiers!) But if we have at our disposal a mathematical system S that includes the arithmetic of the natural numbers plus some set theory, things are considerably simplified. For then we can take, as an abstract counterpart of the first premise, the claim

1'. The cardinality of the set of aardvarks is 21;

1' is an abstract counterpart of 1 because the equivalence of 1' and 1 is provable in N + S.[17] Abstract counterparts of the other premises, and of the conclusion, are as follows:

2'. All sets in the range of the function whose domain is the set of aardvarks, and which assigns to each entity in its domain the set of bugs on that entity, have cardinality 3.
3'. The function mentioned in 2' is 1–1 and its range forms a partition of the set of all bugs.
4'. The cardinality of the set of all bugs is 63.

But now in S we can prove:

(a) If all members of a partition of a set X have cardinality α, and the cardinality of the set of members of the partition is β, then the cardinality of X is $\alpha \cdot \beta$.
(b) The range and domain of a 1–1 function have the same cardinality; and
(c) $3 \cdot 21 = 63$.

But 1', 2', and 3', in conjunction with (a)–(c), entail 4'; and since 1'–4' are abstract counterparts of 1–4, i.e. their equivalence with 1–4 is provable in N + S, we have proved 4 from 1–3 in N + S. So, by Principle C, 4 must follow from 1–3 in N alone.

It is by some argument such as this that we know that 4 follows from 1–3 in N; certainly it isn't on the basis of having gone through a derivation in N that we know this.

---

[17] To simplify things I haven't shifted from N to N* in this case, because in this example such a shift isn't needed. If we did shift from N to N*, we would rewrite 1 as

1* There are exactly twenty-one aardvarks that are not mathematical objects.

and take as an abstract counterpart of 1* the claim

(1*)' The cardinality of the set of aardvarks that are not mathematical objects is 21.

This illustration[18] of the application of mathematics is a very special one. Its special nature is illustrated by the fact that nothing was assumed about the theory N other than that it contained the logic of identity (supplemented with the numerical quantifiers; but these are in principle superfluous). This is not typical of the application of abstract entities in general, though it is typical of the application of the arithmetic of natural numbers. The fact that the natural numbers can find useful application outside the context of any powerful and specialized theories is what is behind the widely shared feeling that the arithmetic of natural numbers has a very special epistemological place. (Cf. for instance Kronecker's remark "God created the natural numbers, all the rest is the work of man.")

But the fact that the arithmetic of natural numbers has this special status is not sufficient grounds to grant that it is true. For I have explained its special status instrumentally: its special status arises from its utility, and since we've shown that it is always in principle eliminable (i.e. you don't get any results with it that you couldn't get without it), its utility is no grounds for believing it true.

---

[18] Hilary Putnam gives a similar illustration in Putnam 1967: cf. pp. 26–33, and in particular pp. 31–3, where he points out that the application of number theory requires only the consistency of mathematics. I was in fact originally led to the view that I take in this monograph largely by thinking about these striking remarks of Putnam's. Note, however, that the conclusion that Putnam draws from his remarks is rather different from the one I draw: his conclusion is that we should interpret pure mathematics as asserting the possible existence of physical structures satisfying the mathematical axioms, whereas my conclusion is that we don't need to interpret pure mathematics at all.

In Putnam 1975 he takes back the view put forth in the earlier paper, claiming in effect that the account given of the application of number theory couldn't possibly be extended to an account of how the theory of functions of real variables is applied to physical magnitudes. (Cf. 1975: 74–5. Putnam has presented this point at greater length in Putnam 1971.) Perhaps in part his pessimism is due to the assumption that any extension of the account of how number theory is applied would have to be put into the framework of a reinterpretation of mathematics; in any case, the later chapters of this monograph (starting with Chapter 3) show how to perform the extension in question, if we forget about reinterpreting pure mathematics and worry only about reinterpreting its applications.

3

Second Illustration of Why Mathematical Entities are Useful: Geometry and Distance

Let us turn now to more complicated applications of abstract entities. Here, too, the situation fits the general description given in the second paragraph of Chapter 2: abstract entities are useful because we can use them to formulate abstract counterparts of concrete statements; then in proving a conclusion in N* from premises in N*, we can at any convenient point 'ascend' from concrete statements to their abstract counterparts, proceed at the abstract level for a while, and then finally 'descend' back to the concrete.

In the cases of application of mathematics that I will now turn to—which are the cases most important for physical theory—the key to carrying out the general strategy of finding 'abstract counterparts' is proving a *representation theorem*. Suppose that using some mathematical theory S which satisfies Principle C of Chapter 1, we can prove the existence of some mathematical structure $\mathfrak{B}$ with certain specified properties. If, using N* + S, we can then prove the existence of one or more homomorphisms (structure-preserving mappings) from concrete objects (or k-tuples of concrete objects) into that mathematical structure $\mathfrak{B}$, then such a homomorphism will serve as a 'bridge' by which we can find abstract counterparts of concrete statements. Consequently, premises about the concrete can be 'translated into' abstract counterparts; then, by reasoning within S, we can prove abstract counterparts of further concrete statements, and then use the homomorphism to descend to the concrete statements of which they are abstract counterparts. The concrete conclusions so reached would always be obtainable without the ascent into the abstract (provided that the mathematical theory S satisfies Principle C); but the ascent into the abstract is often a tremendous saving of time and effort.

Let me illustrate this with an example: Hilbert's axiomatization of Euclidean geometry (Hilbert 1905). Any fully formulated physical theory will include a theory of physical space (or better, of space-time; but since our concern for the moment will be with Euclidean geometry, let's just consider space). Euclidean geometry, considered as a theory of physical space (which is how Euclid originally conceived it) is actually false, but that doesn't matter for my purposes: a false theory is still a theory, and we can use such a theory to illustrate the applicability of mathematical systems like the system of real numbers. Hilbert's formulation of the Euclidean theory is of special interest here because (besides being rigorously axiomatized) it does not employ the real numbers in the axioms; nevertheless, it explains why the system of real numbers can be usefully applied in geometric reasoning.

Without purporting to be very precise, we can say that Hilbert's theory is one in which the quantifiers range over regions of physical space, but do not range over numbers. The predicates of the theory include several, such as 'is a point', which need not concern us. In addition they include the following:

(a) a three-place predicate *between*, where 'y is between x and z' (symbolically, 'y Bet xz') is understood intuitively to mean that y is a point on the line-segment whose endpoints are x and z (the case where $y = x$ or $y = z$ is allowed, i.e. we're dealing with what I'll call *inclusive* betweenness);

(b) a four-place predicate of *segment-congruence*, which I'll write as 'xy Cong zw', understood intuitively to mean that the distance from point x to point y is the same as the distance from point z to point w;

and perhaps also

(c) a six-place predicate of *angle-congruence*, which I'll write as 'xyz A-Cong tuv', understood intuitively to mean that the angle formed by points x, y, and z with vertex at y is the same size as the angle formed by points to t, u, and v with vertex at u.

(The last of these predicates doesn't actually need to be taken as primitive, it can be defined in terms of the others.) Now, I have explained (b) and (c) intuitively in terms of distance and angle-size. But these explanations are not part of the theory: in fact the notions of distance and angle-size can't be defined in the theory (as is obvious from the fact that the theory doesn't quantify over real numbers).

The fact that these quantitative notions are not definable in the theory might appear to raise a problem for Hilbert's formulation, for much of the reasoning in a typical book on Euclidean geometry proceeds in terms of the lengths of line-segments and/or the size of angles: in fact, many of the theorems are explicitly theorems about lengths (e.g. Pythagoras's theorem). Does this mean that Hilbert left something out? No, for he proved the kind of theorem I'm calling a *representation theorem*: he proved (in a broader mathematical theory) that given any model of the axiom system for space that he had laid down, there would be at least one function d mapping pairs of points onto the non-negative real numbers, satisfying the following 'homomorphism conditions':

(a) for any points x, y, z and w, xy Cong zw if and only if $d(x,y) = d(z,w)$;
(b) for any points x, y and z, y is between x and z if and only if $d(x,y) + d(y,z) = d(x,z)$.

So if we take d to represent distance, segment-congruence becomes 'equivalent' to just the claim about distance we'd expect, and similarly for betweenness. (Hilbert also proved the existence of a function m mapping triples of points into numbers, satisfying analogous conditions: m was a representation for angle-sizes.) Given these results it was easy to show that the standard Euclidean theorems about lengths and angle-sizes would be true if restated as theorems about any functions d and m meeting the given conditions. So in the geometry itself we can't talk about numbers, and hence we can't talk about distances or angle-sizes; but we have a metatheoretic proof which associates claims about distances or angle-sizes with what we can say in the theory. Numerical claims then, are abstract counterparts of purely geometric claims, and the equivalence of the abstract counterpart with what it is an abstract counterpart of is established in the broader mathematical theory.

Incidentally, in addition to the representation theorems Hilbert established *uniqueness theorems*, one for distance, one for angle-size: e.g. the uniqueness theorem for distance says that if d_1 and d_2 are two functions mapping pairs of points into non-negative reals, both of which satisfy the two conditions just laid down, then d_1 and d_2 differ only by a positive multiplicative constant; and conversely, that if d_1 and d_2 differ only by a positive multiplicative constant, then d_1 satisfies (a) and (b) if and only if d_2 does. Thus the fact that geometric laws, when formulated in terms of distance, are invariant under multiplication of all distances by a positive constant, but are not invariant under any other transformation of scale, receives a satisfying *explanation*: it is explained by the *intrinsic facts* about physical space, i.e. by the facts about physical space which are laid down without reference to numbers in Hilbert's axioms. This is a point that will be important later, but for now let's go back to the representation theorem.

Hilbert's representation theorem, I've said, shows that statements that talk about space alone, without reference to numbers, are equivalent to certain 'abstract counterparts' which do talk about numbers. Because of this, we can use the theorem as a device for drawing conclusions about space (conclusions *statable without* real numbers) much more easily than we could draw them by a direct proof from Hilbert's axioms. For instance, it is not difficult to say intrinsically (see Figure 1):

(a) that a_1, a_2, a_3 and b_1, b_2, b_3 form right triangles with right angles at a_2 and b_2;
(b) that there is a segment cd such that $a_1 a_2$ is twice the length of cd, $a_2 a_3$ is five times the length of cd, $b_1 b_2$ is three times the length of cd, and $b_2 b_3$ is four times the length of cd. (E.g. we say that $a_1 a_2$ is twice the length of cd by saying that there is a point x between a_1 and a_2 such that $a_1 x$ Cong cd and $x a_2$ Cong cd.)

One might then wonder whether $a_1 a_3$ is longer than $b_1 b_3$. If one tries to answer this without using the representation theorem, it will be very difficult. But if one uses the representation theorem, one can invoke Pythagoras's theorem to quickly establish that $a_1 a_3$ is $\sqrt{29}$ times the length of cd and that $b_1 b_3$ is five times the length of cd and therefore that $a_1 a_3$ is indeed longer than $b_1 b_3$.

So invoking real numbers (plus a bit of set theory) allows us to make inferences among claims not mentioning real numbers much more quickly than we could make those inferences without invoking the reals. And the inferences we make in this way will be correct every time.

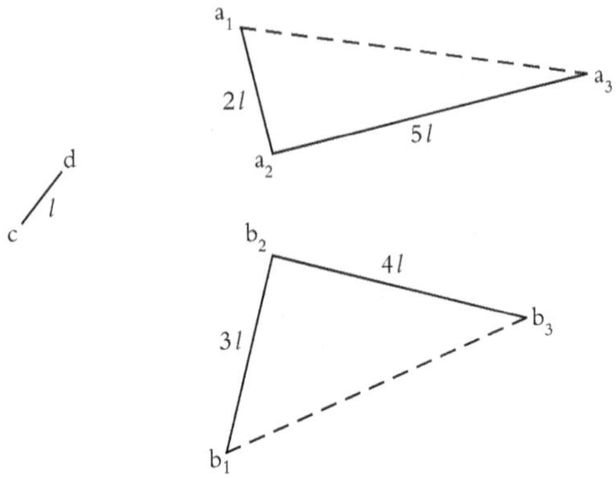

30 GEOMETRY AND DISTANCE

Figure 1

Prima facie, this might seem to be good evidence that the theory of real numbers (plus some set theory) is true: after all, if it weren't true, invoking it in arguments in this way ought to sometimes lead from otherwise true premises to a false conclusion. But we've seen in Chapter 1 that this *prima facie* plausible argument is deeply mistaken: the fact that the theory of real numbers (plus set theory) has this truth-preserving property is a fact that can be explained without assuming that it is *true*, but merely by assuming that it is *conservative*, which is a different matter entirely; in fact, as remarked in the Appendix, we really need only to assume a restricted form of conservativeness, which follows from the *consistency* of set theory alone.

4
Nominalism and the Structure of Physical Space

The reader might reasonably wonder about the assertion at the very end of the previous chapter: after all, Principle C says that when mathematical theories are added to nominalistic theories, you can never deduce any nominalistic consequences you couldn't deduce otherwise; but I haven't yet claimed that Hilbert's formulation of the Euclidean theory of space is genuinely nominalistic, I have claimed only that it doesn't quantify over *real numbers*. Now, this worry can be easily alleviated: for whether or not Hilbert's theory ought to be counted nominalistic on philosophical grounds, there can be no doubt that (if set theory is consistent) our mathematical theories apply to it in a conservative fashion. I will explain this, but first I want to raise the more controversial question of whether Hilbert's formulation of the Euclidean theory of physical space *can* be counted as genuinely nominalistic on philosophical grounds. This question raises several important issues.

I

Some of these issues can be brought out by considering the following objection. 'Hilbert's axiomatization of geometry just builds into physical space all the complexity and structure that the platonist builds into the real number system. For instance, Hilbert's axiomatization requires physical space to be uncountable, and in fact requires lines in physical space to be isomorphic to the real numbers. And there doesn't seem to be a very significant difference between postulating such a rich physical space and postulating the real numbers.'

In reply to this, let me first remind the reader that as I am conceiving nominalism, the nominalistic objection to using real numbers was not on the grounds of their uncountability or of the structural assumptions (e.g. Dedekind completeness) typically made about them. Rather, the objection was to their abstractness: even postulating one real number would have been a violation of nominalism as I'm conceiving it.[19] Conversely, postulating uncountably many *physical* entities (e.g. uncountably many parts of a physical object, or of a light ray, or, as here, of physical space itself) is not an objection to nominalism; nor does it become any more objectionable when one postulates that these physical entities obey structural assumptions analogous to the ones that platonists postulate for the real numbers.

Perhaps it is a bit odd to use the phrase 'physical entity' to apply to space-time points.[20] But however this may be, space-time points are not abstract entities in any normal sense. After all, from a typical platonist perspective, our knowledge of mathematical structures of abstract entities (e.g. the mathematical structure of real numbers) is *a priori*; but the structure of physical space is an empirical matter. That is, most platonists who believe current physical theory believe that it is *a priori* true that there are real numbers obeying the usual laws, and that it is a high-level empirical hypothesis (not easily disconfirmed, but subject to revision by the development of an alternative physical theory) that there are lines in space which (locally anyway) are isomorphic to the real numbers. No platonist would identify the real numbers with the points on any physical line: for one thing, it would be arbitrary which such line one picked to identify the real numbers with, and arbitrary which point on the line to identify with 0 and which with 1; but more fundamentally, to make any such identification would be to identify the real numbers with something we can know about only empirically. (Occasionally it is suggested by those seeking a satisfactory formulation of quantum mechanics that we ought to view space and time as quantized. To my knowledge, no such proposal has ever been worked out very far; but if one were, and if it turned out to make the best sense of the evidence and best solve the interpretational difficulties of quantum theory, we would have strong empirical reasons

---

[19] *Added in 2nd edition:* I thought it was obvious that this was a joke: the suggestion that we might posit a single object and regard it as a real number is ludicrous. It's bad form to explain when one is joking, but I've been dismayed that several good philosophers have taken the remark at face value.

[20] For the reader who wonders why I say 'space-time point' instead of 'point of space': your curiosity will be alleviated in the last paragraph of section I of this chapter.

to believe that between any two space-time points there are only finitely many others. Surely however we ought not to count such a development as an empirical discovery that there are only finitely many real numbers between 0 and 1.)

Even ignoring these points, there is a further reason that postulating physical space isn't like postulating real numbers: and that is that the ideology that goes with the postulate of points of space is less rich than that which goes with the postulate of the real numbers. With the postulate of real numbers goes the operations of addition and multiplication: no such operations are directly defined on space-time points in Hilbert's theory; indeed none are even implicitly definable since any introduction of an addition or multiplication function on space-time points would have to rely on an arbitrary choice of one point to serve as 0 and another to serve as 1. Something like addition can be reconstructed within Hilbert's theory, but it is addition of intervals rather than of points (and it doesn't give an addition function but rather a non-functional relation, 'interval x is the same length as the sum of intervals y and z'). With multiplication, we can't even reconstruct the relation of one interval being the product of two others: any introduction of such a product relation on intervals would have to depend on an arbitrary choice of one interval to serve as 'the unit interval', and no such notion is employed in the Hilbert theory. The best one can do with the Hilbert primitives is to reconstruct comparisons of products of intervals, and it takes quite a bit of work to reconstruct such comparisons in a suitably generalizable way.[21] These observations make it clear that the objection that we are using the space-time points as if they were real numbers is quite erroneous.

These points are further reinforced by the fact that the usual theory of real numbers includes not only the first-order theory that invokes only the functions of addition and multiplication: it includes also the apparatus of quantification over functions defined on the real numbers, and also enough higher-order sets to enable us to define the continuity, differentiability, etc. of such functions. No such apparatus is invoked in the theory that takes space-time points as the objects of quantification: though we will eventually see that the invariant content of many

---

[21] 'In a suitably generalizable way' means 'in a way generalizable to products of spatio-temporal intervals with scalar intervals'. The suitably generalized way of making product comparisons is given in Chapter 8.

statements of continuity, differentiability, etc. of functions is expressible in the system to be developed, it is to be expressed without referring to or quantifying over functions or anything like functions.

One might think that if the system of space-time points was as distinct from the system of real numbers as I've been saying, then it would be a remarkable coincidence that points on a physical line should happen to have precisely the structure of such an important mathematical system as the real numbers, and that important mathematical operations (e.g. differentiation) on functions of real numbers should have analogs which play an important role in the physical theory. Surely, it could be argued, this can't be a coincidence: doesn't this show then that the physical theory is really platonism in disguise?

The trouble with this objection is that it completely ignores history: the theory of real numbers, and the theory of differentiation etc. of functions of real numbers, was developed precisely in order to deal with physical space and physical time and various theories in which space and/or time play an important role, such as Newtonian mechanics. Indeed, the reason that the real number system and the associated theory of differentiation etc. is so important mathematically is precisely that so many of the problems to which we want to apply mathematics involve space and/or time. It is hardly surprising that mathematical theories developed in order to apply to space and time should postulate mathematical structures with some strong structural similarities to the physical structures of space and time. It is a clear case of putting the cart before the horse to conclude from this that what I've called the physical structure of space and time is really mathematical structure in disguise.

So in summary: there is indeed a good deal in common between on the one hand the structure of physical space that both I and the platonists postulate and on the other hand the structure of mathematical objects postulated by platonists; and there is an obvious reason why there should be this commonality of structure, given that the mathematics was developed to deal with physical space (and time). Still, there are many ways in which the physical structure is less rich than the mathematical structure (e.g. no addition relation defined on points; no multiplication relation defined on points or even on intervals; no functions, sets of functions, etc.). And the physical structure is all an empirical postulate, subject to revision by experience in a way that mathematics is not.

There are, to be sure, certain views of space-time according to which the quantification over space-time points or space-time regions really would be a violation of nominalism. I'm speaking of relationalist views of space-time, as opposed to the substantivalist view. According to the substantivalist view, which I accept, space-time points (and/or space-time regions) are entities that exist in their own right. In contrast to this are two forms of relationalist view. According to the first (*reductive relationalism*), points and regions of space-time are some sort of set-theoretic construction out of physical objects and their parts; according to the second (*eliminative relationalism*), it is illegitimate to quantify over points and regions of space-time at all.[22] It is clear that reductive relationalism is unavailable to the nominalist: for according to that form of relationalism, points and regions of space-time are mathematical entities, and hence entities that the nominalist has to reject. So a nominalist must either be a substantivalist or be an eliminative relationalist, and only in the first case can he find Hilbert's theory acceptable.

It is my view however that independently of nominalism, a substantivalist view is preferable to either form of relationalist view, for a number of reasons most of which cannot be discussed here. I will merely say that I don't think that any relationalist programme, of either a reductive or an eliminative sort, has ever been satisfactorily carried out, even given a full-blown platonistic apparatus of sets. The problem for relationalism is *especially* acute in the context of physical theories that take the notion of a *field* seriously, e.g. classical electromagnetic theory. From the platonistic point of view, a field is usually described as an assignment of some property, or some number or vector or tensor, to each point of space-time; obviously this assumes that there are space-time points, so a relationalist is going to have to either avoid postulating fields (a hard road to take in modern physics, I believe) or else come up with some very different way of describing them. The only alternative way of describing fields that I know is the one I use later in the monograph in connection with the gravitational potential field in Newtonian mechanics: it does without

---

[22] Or anyway, it is illegitimate to quantify over *unoccupied* points and regions: quantification over occupied points or regions (i.e. points or regions wholly occupied by parts of physical objects) could be regarded as equivalent to quantifying over the objects which occupy them, and hence as unproblematic for the relationalist.

the properties or the numbers or vectors or tensors, but it does not do without the space-time points.[23] In general, it seems to me that recent developments in both philosophy and physics have made substantivalism a much more attractive position than it once was; it certainly has been adopted by the majority of the 'new wave' of space-time theorists. (For two good discussions, see Earman 1970 and Friedman 1981.) In any case, substantivalist views of space-time are certainly possible, and on such a substantivalist view it is perfectly nominalistic to quantify over space-time points and/or space-time regions.

Actually this doesn't justify quantifying over points or regions *of space*, if a point or region of space is construed as an entity that endures through time. And indeed, there are real difficulties about quantifying over points or regions of space on any such construal, for on such a construal it would seem to make objective sense to ask whether two non-simultaneous events are at the same point of space, and hence to ask whether a given object is at absolute rest. The notion of absolute rest is one that positivists have quite rightly objected to, in my view: this is a point I will return to briefly in the next chapter. Fortunately, however there is a way to construe quantification over points and regions of space so that it involves no commitment to absolute rest, in any physical theory in which a notion of simultaneity is available: simply regard a claim about space as an abbreviation for the assertion that the claim holds for each of the spatial slices of space-time (i.e. the slices generated by the simultaneity relation). So the claim that physical space is Euclidean is translated into the claim that each of the spatial slices of space-time is Euclidean. It is trivial to rewrite Hilbert's axiomatization of the geometry of space so that that is explicitly what it says; if we do so, then the objects in the domain of the quantifier are really space-time points rather than points of space, and there can be no danger of viewing the theory as being committed to the idea that absolute rest is a physically significant notion. (I won't bother to explain

---

[23] Note incidentally that according to theories that take the notion of a field seriously, space-time points or regions are full-fledged causal agents. In electromagnetic theory for instance, the behavior of matter is causally explained by the electromagnetic field values at unoccupied regions of space-time; and since, platonistically speaking, a field is simply an assignment of properties to points or regions of space-time, this means that the behavior of matter is causally explained by the electromagnetic properties of unoccupied regions. So according to such theories space-time points are causal agents in the same sense that physical objects are: an alteration of their properties leads to different causal consequences.

how to rewrite Hilbert's theory in this way however, since the theory that resulted would be of less use than a stronger nominalistic theory about space-time structure to be set out in Chapter 6.)

II

I have allowed our nominalist to quantify over points or regions of space-time. Is there any reason why he shouldn't quantify over both points and regions? Some philosophers would be willing to accept the existence of certain kinds of regions—say, regular open regions—but not of points. This is not a view I *object to*: it may well be possible to find nominalistic systems similar in many respects to the Hilbert system (and to the systems to follow later on in the book), but that quantify over arbitrarily small regular open regions instead of over points; and if it *is* possible, then the nominalist has no reason to object to dispensing with points in favor of regular open regions. But I also do not see that the nominalist has any particular *reason* to forego points for arbitrarily small regular open regions—the desire for such purity is a quasi-finitist desire, not a nominalist desire. Since the desire to forego points is not one I share, and since it appears to be mathematically difficult, I will make no attempt to satisfy that desire in this book.

How about the converse question: given a nominalism in which we quantify over space-time points, is there any added difficulty in quantifying over regions? If our nominalist accepts Goodman's calculus of individuals (Goodman 1972: part IV), then the introduction of points carries with it the introduction of regions: for a region is just a *sum* (in Goodman's sense) of the points it contains.[24] And even if one does not accept the calculus of individuals in general—even if one thinks that there are entities that can't meaningfully be 'summed'—there seems to be little motivation for allowing points and yet disallowing regions: in fact, it seems attractive to regard points of space-time as a special case of regions, namely as regions of minimal size. So it seems to me that regions are nominalistically acceptable. (I should note however that only fairly

---

[24] As the reference to Goodman indicates, I use 'region' in such a way that there is no empty region, i.e. no region containing no space-time points. Also regions don't need to be connected, or measurable, or anything like that: very 'unnatural' collections of points count as regions.

'regular' regions are directly used in the monograph, so a nominalist who would balk at the use of highly 'irregular' regions need not balk at the uses to which regions will actually be put.)[25]

If these claims about what should count as nominalistic are accepted, then there is at least an important sense in which Hilbert's formulation of the Euclidean theory of space is nominalistic, or can be made so with a little rewriting. Hilbert's theory is usually formulated as a second-order theory, in which the first-order variables range over points, lines, and planes; in other words, the first-order variables range over regions of various kinds. Consequently, the second-order variables range over *sets* of points, lines, and planes, and that doesn't look very nominalistic. However, only one second-order axiom is really needed, the Dedekind continuity axiom; and this axiom quantifies only over non-empty sets of *points*. This is important, for in the absence of any further use of sets, there is no substantive difference between a non-empty *set* of points on the one hand and a Goodmanian sum of points, or a region, on the other. So we can regard the second-order quantifiers in Hilbert's theory as ranging over regions. (And if we like, we can then restrict the range of the first-order quantifiers to points, either by using second-order quantifiers whenever we want to speak of lines and planes, or by paraphrasing claims about lines and planes in terms of claims about points and the relation of betweenness.) If we write Hilbert's theory in this way, then the quantifiers (both first-order and second-order) range only over regions of space; and I've argued that regions of space are nominalistically acceptable entities. *So if we write Hilbert's formulation of the Euclidean theory of space in this way, it has a purely nominalistic ontology.*

It does, admittedly, have a logic that one might find objectionable: it involves what might be called *the complete logic of the part/whole relation*, or *the complete logic of Goodmanian sums*, and this is not a recursively axiomatizable logic. To clarify this, note that the theory as I've suggested it be written is still a second-order theory, that is, it still involves second-order logic: it is merely that because of the nature of the objects in the range of the first-order quantifiers (viz. because they do not overlap), and because also we haven't invoked variables for functions or for

---

[25] This is not to deny that there might be difficulties in figuring out how to axiomatize the 'regular' regions without assuming the existence of the 'irregular' ones. How difficult this task would be presumably depends on the concept of regularity involved.

predicates of more than one place, no nominalistically dubious entities need be invoked to serve in the range of the second-order quantifiers. This ontological difference is perhaps sufficiently striking so that we ought not to call the logic 'second-order logic' anymore, but something else, such as 'the complete logic of Goodmanian sums'; nonetheless, the consequence relation is still like that of second-order logic, which is not recursively axiomatizable. Consequently, insofar as one objects to the strength of the second-order consequence relation, one will object to this version of Hilbert's formulation of the Euclidean theory of space.

I share the feeling that the invocation of anything like a second-order consequence relation is distasteful, and will discuss the possibility of eliminating it in the final chapter of the book. For now, let me simply note that for platonistic theories too, the most natural and intuitive formulation of a theory is often a second-order formulation. For instance, intuitive set theory—by which I mean not the intuitive Cantorian set theory that was shown inconsistent, but the intuitive set theory that underlies the Zermelo-Fraenkel and similar axiomatizations—is a second-order theory: e.g. it will include as an axiom or a theorem the second-order separation principle

$\forall P \forall x \exists y \forall z (z \in y \leftrightarrow z \in x \wedge P(z))$.

To get a first-order axiomatization we have to weaken the theory, replacing the second-order axiom or axioms by schemas of first-order axioms, namely the schema of replacement and/or separation. This first-order weakening of intuitive set theory has a lot of 'non-standard' models (e.g. models in which sets that are really infinite satisfy the formula that is usually regarded as defining finiteness): such models are 'non-standard' precisely because they are not models of second-order set theory.[26] Similarly, the second-order Hilbert axiomatization of geometry can be weakened to a first-order system, in either of two ways: a severe

---

[26] See Montague 1965: 131–48, for the sort of second-order axiomatization I have in mind, and a defense of the idea that not only in set theory but elsewhere as well, the way to explicate the idea of a standard model of a first-order theory is as 'model of an associated second-order theory'. As Montague points out, the models of Zermelo-Fraenkel set theory that are 'standard' on this explication are precisely those models that are isomorphic to models in which the domain is the set of all sets of rank less than α for some strongly inaccessible α (greater than ω), and in which '$\in$' is assigned the membership relation restricted to this domain. I agree with Montague that this is the most natural notion of a standard model for set theory.

weakening which entirely drops the use of regions bigger than points has been studied by Tarski (1959), and a less severe weakening to a first-order axiomatization will be mentioned in the final chapter. But these first-order weakenings of the Hilbert system all have non-standard models. These non-standard models together with the non-standard models of first-order set theory make the question of the relation between the first-order nominalistic theory and the first-order platonistic theory harder to settle; a representation theorem like Hilbert's is much easier to state and prove if it is taken as relating the intuitive (second-order) nominalistic geometry to the intuitive (second-order) set theory than if it is taken as relating their first-order weakenings. For this reason I will put off the issue of first-order axiomatization until the final chapter.

Since I am putting that off, it is necessary to make sure that nothing in my remarks in the previous chapter, about the philosophical significance of Hilbert's representation theorem, turned on the false assumption that Hilbert's axiomatization was first order. The only remark which might seem suspect from this point of view came at the very end of the chapter. After pointing out that mathematical entities (real numbers together with functions from space-time points into the reals) can usefully be employed in connection with Hilbert's axiomatization, and that when they are employed we are never led to a false conclusion about space from true premises, I raised the question of whether this fact is evidence that the theories which postulate mathematical entities are true. My answer was no: we could, I claimed, explain the truth-preservingness of mathematics in this context entirely by its conservativeness, which is a much weaker (or more accurately, a quite different) property; in fact, I remarked that we really only need to assume a restricted form of conservativeness, which follows from the consistency of mathematics alone. This, however, raises a question: is the consistency of mathematics (i.e. the consistency of set theory, since mathematics reduces to set theory) sufficient to entail that mathematics can be employed in reasoning about second-order theories in a truth-preserving way? The answer is that the semantic consistency of *second-order* set theory *is* sufficient for this conclusion: in fact, the main arguments of the Appendix to Chapter 1 go over with little alteration when all the theories are taken to be second order.[27] The upshot is that in the

---

[27] In more detail: recall that conservativeness as I defined it initially is a *semantic* notion, one involving *consequence* rather than *provability*. In the Appendix to Chapter 1,

context of reasoning about Euclidean geometry at least, the nominalist can invoke the theory of real numbers (with the attendant functions) as much as he likes, for he is guaranteed that he can never be led into error by so doing.

I reformulated it in terms of consistency; this is ambiguous between the semantic and the syntactic, but in referring to some of the arguments as proof-theoretic, and in the way I wrote the proof in note 15, I showed that it was the syntactic notion I was dealing with. The justification for the shift from semantic to syntactic notions is of course the Gödel completeness theorem for first-order logic. In the case of second-order logic there can be no such completeness theorem: here, we must stick to semantic notions throughout. But the key results of the Appendix remain unchanged. In particular, if 'consistent' in (C_0) is understood as 'semantically consistent', the set-theoretic proof of (C_0) is as before: the method described for turning a model of T into a model of $ZFU_{V(T)} + T^*$ can remain unchanged as long as both ZFU and T are second-order theories. (Recall the remarks in note 26 on what the models of second-order set theory are like.) Analogously, the proof in note 15 that (C_1) follows from the consistency of ZF needs no alteration when T and ZF are made second order, except that since we're replacing syntactic consistency by semantic consistency, the step involving the Gödel completeness theorem is unnecessary. (Two less central results of the Appendix are more problematic: the proofs via the Robinson theorem (which is not valid in second-order logic) and Weinstein's proof that the ω-consistency of ZF suffices for (C_2). But these results are not required for the remarks in the text to be true.)

5

My Strategy for Nominalizing Physics, and its Advantages

So far, I have not tried to argue that we can come up with nominalistic theories to replace platonistic ones: I have merely argued that *if* we had a nominalistic theory, then it would be legitimate to introduce mathematics as an auxiliary device that aids us in drawing inferences; and I have tried to indicate why that auxiliary device would be useful, and to show that its usefulness as an auxiliary device is no grounds whatever for supposing that it consists of a body of truths. The real question then is whether an *attractive* nominalistic formulation of physics is possible. I say an attractive nominalistic formulation, because if no attractiveness requirement is imposed, nominalization is trivial: simply take as axioms of your physical theory all the nominalistically statable consequences of the platonistic formulation of the theory. (Or, if you want a recursive set of axioms, take the Craigian transcription of the set of nominalistically statable consequences.) Obviously, *such* ways of obtaining nominalistic theories are of no interest. The way that I will suggest of obtaining nominalistic theories is very different from this.

In order initially to motivate the idea that an attractive nominalistic formulation of physics is possible, let us return to Hilbert's axiomatization of geometry. There are two approaches to axiomatizing geometry, sometimes called *the metric approach* and *the synthetic approach*. In the metric approach we take as primitive a particular function-symbol d, which we regard as denoting a particular mapping of pairs of points of space into the real numbers. Then if we regard the mathematical laws of real numbers, functions, and so forth as independently given, we can use d to lay down a relatively simple set of axioms for the geometry. The synthetic approach is the one that Hilbert followed, the one which does without real numbers, functions, etc. This approach is also the one that Euclid had (less rigorously) followed long before—Euclid *had to* follow the synthetic approach, because the theory of real numbers hadn't been sufficiently developed in his day for the metric approach to be possible. (The real numbers were in fact first introduced into mathematics as a means of simplifying geometric reasoning.) But to anyone already familiar with the theory of real numbers, the metric approach is a good deal easier, and for that reason it is used in many recent books in geometry. If one were familiar only with the metric approach to Euclidean geometry, one would probably conclude that one needs to quantify over real numbers in developing a theory of the geometry of space. The Hilbert axiomatization, however, shows that this is not so.

My guess is that the same is true for other physical theories. Insofar as they've been rigorously formulated at all, they've been formulated platonistically, for it is easier to formulate a theory that way when one has a sufficiently developed mathematics. My guess, however, is that a thorough foundational analysis of such theories will show that reference to real numbers, etc. is no more necessary in them than it is in geometry. And this isn't a mere guess: I substantiate it in Chapters 6–8 with respect to one physical theory, viz. Newton's theory of gravitation; and it would be routine to extend the nominalistic treatment of gravitational theory to other theories with a similar format, such as special relativistic electromagnetic theory.

I believe that such 'synthetic' approaches to physical theory are advantageous not merely because they are nominalistic, but also because they are in some ways more illuminating than metric approaches: they explain what is going on without appeal to extraneous, causally irrelevant entities. The attempt to eliminate theoretical entities of physics (e.g. electrons) from explanations of observable phenomena is not likely to be possible without bizarre devices like Craigian transcriptions; it is also not likely to be illuminating even if it is possible, because electrons are causally relevant to the phenomena they are invoked to explain. But even on the platonistic assumption that there are numbers, no one thinks that those numbers are causally relevant to the physical phenomena: numbers are supposed to be entities existing somewhere outside of space-time, causally isolated from everything we can observe. If, as at first blush appears to be the case, we need to invoke some real numbers like

6.67×10^{-11} (the gravitational constant in $m^3/kg^{-1}/s^{-2}$) in our explanation of why the moon follows the path that it does, it isn't because we think that that real number plays a role as a cause of the moon's moving that way; it plays a very different role in the explanation than electrons play in the explanation of the workings of electric devices. The role it plays is as an entity *extrinsic to the process to be explained*, an entity related to the process to be explained only by a function (a rather arbitrarily chosen function at that). Surely then it would be illuminating if we could show that a purely intrinsic explanation of the process was possible, an explanation that did not invoke functions to extrinsic and causally irrelevant entities.

In saying that this is an advantage, I don't mean to suggest that extrinsic explanation should always be avoided: the point is rather that from a proper synthetic theory, one will be able to prove the equivalence of the intrinsic and extrinsic explanations. (That is, one will be able to prove that the two explanations are equivalent given the assumption that the entities involved in the extrinsic explanation exist. If one believes that they don't exist, then one will hold that the extrinsic explanation is merely a useful fiction, but one which can be used in good conscience by anyone who knows of the intrinsic explanation, because of the conservativeness of mathematics.) An illustration of this is provided by synthetic geometry: given the axioms of synthetic geometry, one can prove (given standard mathematics) the equivalence of on the one hand explanations of features of physical space stated in terms of betweenness and congruence and on the other hand extrinsic explanations involving quantitative distance and angle measures; hence one is free to use the extrinsic explanations in practice.

I am saying then that not only is it much likelier that we can eliminate numbers from science than electrons (since numbers, unlike electrons, do not enter causally in explanations), but also that it is more illuminating to do so. It is more illuminating because the elimination of numbers, unlike the elimination of electrons, helps us to further a plausible methodological principle: the principle that *underlying every good extrinsic explanation there is an intrinsic explanation*. If this principle is correct, then real numbers (unlike electrons) have got to be eliminable from physical explanations, and the only question is precisely how this is to be done.

Note that the principle I've italicized is not a nominalistic principle: it could perfectly well be accepted by a platonist, though of course, not by any platonist who believed that one could argue for platonism by saying that mathematical entities are needed for physics. Conversely, a nominalist need not accept the principle. There are indeed ways of trying to establish the possibility of nominalism that, even if successful, would not establish the italicized principle. One such approach is that of Charles Chihara 1973. Chihara's approach is one of those alluded to in the introduction, on which one tries to *reinterpret* mathematics: in this case, one reinterprets it as being about linguistic entities instead of abstract entities. I find my approach preferable to his for three reasons. In the first place, as Chihara of course recognizes, the linguistic view requires that only predicative mathematical reasoning be used in application, and it isn't at all obvious that we don't need impredicative reasoning in doing science. (My view licenses (but doesn't demand) the use of impredicative reasoning, as we shall see in Chapter 9.) In the second place, the linguistic entities that Chihara appeals to include sentence types no token of which has even been uttered, and it is not at all obvious to me whether these should count as nominalistically legitimate. But third and most fundamental, Chihara's view does nothing to illuminate the use of extrinsic, causally irrelevant entities in the application of mathematics. That is, Chihara's methods do not show us how to provide intrinsic explanations underlying extrinsic explanations; they merely show that linguistic surrogates of mathematical entities can be used in place of mathematical entities in our extrinsic explanations (a fact which I take to be uninteresting, since as I've argued, there is no need in the mathematical case to regard extrinsic explanations as literally true).

I conclude this chapter by noting that one of the things that gives plausibility to the idea that extrinsic explanations are unsatisfactory if taken as *ultimate* explanation is that the functions invoked in many extrinsic explanations are so arbitrary. For example, in the case of geometry, the choice of one distance function over any other one which differs from it by positive multiplicative constant is completely arbitrary; it reflects in effect an arbitrary choice of units for distance. (When we move from geometry to physics generally, there is in the metric approach not only an arbitrary choice of a unit of distance, but also an arbitrary choice of units for other quantities, an arbitrary choice of a rest frame, and various other arbitrary choices as well.) Now an analogous arbitrariness *could* exist on an intrinsic approach too: it would exist if we singled out a particular pair of points of space-time (say, the endpoints of a certain platinum rod in the Bureau of Standards at such and such a time), and constantly referred to this pair of points in making distance comparisons when we developed the theory. Hilbert, however, did not resort to such an unaesthetic move in his intrinsic development of geometry; nor shall I resort to it in my intrinsic formulation of gravitational theory. What Hilbert did do (in his uniqueness theorem) was to *explain, in terms of intrinsic facts about space which are statable without such arbitrary choices, why the choice of functions to be invoked in the extrinsic theory will be arbitrary to precisely the extent that it is.* This feature of the Hilbert approach to geometry is highly attractive, and it is a feature I will take pains to emulate when I extend the synthetic treatment of geometry to a synthetic treatment of gravitational theory.

6
A Nominalistic Treatment of Newtonian Space-Time

I turn now to the problem of giving a nominalistic formulation of physics, a formulation which meets the additional constraints imposed in Chapter 5: it is to be 'attractive', unlike Craigian axiomatizations; it is to be a 'purely intrinsic' formulation; and it is to be a formulation that does not appeal to arbitrarily chosen objects to serve as units of length, or to arbitrarily chosen systems of coordinates, or to any such thing. These further constraints are not very precise, but I hope that they are *reasonably* clear; for I will implicitly and sometimes explicitly invoke these constraints (especially the last one) in motivating the construction to follow.

The first step in giving a nominalistic formulation of physics is to give a nominalistic treatment of space-time. I've already discussed a nominalistic treatment of space, but space-time is a little different, both in Newtonian mechanics and in special relativity. It is different not just in being 4-dimensional instead of 3-dimensional, but in not having a full Euclidean structure. (Also in having some extra structure not present in Euclidean 4-space.)

In the Newtonian case, the lack of a full Euclidean structure comes out in two ways. First, there is no 'objective' way to compare spatial distance with temporal distance; that is, although one could arbitrarily define such a comparison (e.g. by saying that the spatial distance between two points was equal to the temporal distance if the temporal distance was the same as was required for a certain uniformly moving object in the Bureau of Standards to traverse that spatial distance), nonetheless there is no one such means of comparison that is naturally singled out by the laws of Newtonian mechanics.

In order to explain the second way in which space-time lacks full Euclidean structure, I must digress to discuss the issue of absolute rest.

As I said in Chapter 4, I do not think that the notion of rest makes objective sense in Newtonian mechanics: it makes sense only relative to an arbitrary choice of coordinate system. Newton himself disagreed with this conclusion: he thought that the notion of absolute rest (i.e. rest that isn't *merely* rest relative to a coordinate system) was required in order to formulate the laws of mechanics, and consequently that it must make objective sense. In support of the idea that the concept was needed to formulate the laws of mechanics Newton produced his famous bucket argument: this argument makes a strong case for the idea that you need a notion of absolute *acceleration* to formulate the laws of mechanics, and Newton thought that the only way to explain absolute acceleration was in terms of absolute velocity, so that that must make objective sense too. (And if absolute velocity makes objective sense, so of course does absolute rest: something is at absolute rest if its absolute velocity is zero.) This is certainly a persuasive argument for the claim that talk of absolute rest makes objective physical sense. Nevertheless the conclusion of the argument is undeniably embarrassing, for no one of the unaccelerated frames is naturally singled out as the rest frame by the laws of the theory.

How are we to get around this embarrassing conclusion? One way of course would be to change the physical theory, so that absolute acceleration is not used: this was Mach's program. But there is an alternative move which does not (in any very significant sense anyway) change the physical theory, and that is to give a treatment of absolute acceleration which doesn't take it as defined in terms of a numerical velocity; this allows us to have absolute acceleration without absolute rest. A platonistic treatment of acceleration (using 4-dimensional tensor methods) that accords with this idea is known, and is now popular with philosophers of physics. (Cf. for instance the papers by Earman and Friedman cited in the previous chapter.) In committing myself to the avoidance of arbitrary choices, I committed myself to coming up with a nominalistic treatment with the same virtue. (As I will later remark, I think that the nominalistic approach does even better than the tensor approach in avoiding arbitrary choices.)

What does all this have to do with the structure of space-time? I said above that there is no 'objective' way to compare spatial distance with temporal distance. We can now see that in Newtonian mechanics there is not even an 'objective' way to compare the spatial distance between space-time points x and y with the spatial distance between z and w, except in the case where x is simultaneous with y and z is simultaneous with w. To assume a more general comparison of spatial distance is to assume a notion of sameness of place across time, i.e. a notion of absolute rest; and this notion makes no objective sense in Newtonian mechanics. (Again, sense can be given to the notion of rest, by arbitrary stipulation of a rest frame; but no one rest frame is naturally singled out by the laws of the theory.)

We've seen two respects in which space-time lacks full Euclidean structure. We must bear these facts in mind in trying to give an intrinsic treatment of the geometry of space-time: we must describe space-time geometry intrinsically without attributing to space-time any structure that isn't objectively there.

It turns out to be quite easy to give an intrinsic account of the geometry of space-time, both for Newtonian mechanics and for special relativity, by building on an intrinsic treatment of *affine* geometry that has been provided by Szczerba and Tarski 1965.

Before discussing this, let me first return to Hilbert's representation and uniqueness theorems. Hilbert actually proved more general representation and uniqueness theorems than the ones stated in Chapter 3: his theorems, in their full generality, invoke not distance functions, but coordinate functions from which distance functions can be defined. The more general theorems are as follows:

(R_E) A structure $\langle \mathcal{A}, \text{Bet}_\mathcal{A}, \text{Cong}_\mathcal{A} \rangle$ (where $\text{Bet}_\mathcal{A} \subseteq \mathcal{A} \times \mathcal{A} \times \mathcal{A}$ and $\text{Cong}_\mathcal{A} \subseteq \mathcal{A} \times \mathcal{A} \times \mathcal{A} \times \mathcal{A}$) is a model of the Hilbert axioms if and only if there is a 1-1 function Φ from $\mathcal{A}$ onto $\mathbb{R}^3$ (the set of ordered triples of real numbers) such that if we define $d_\Phi(x, y)$ for x,y in $\mathcal{A}$ as

$$\sqrt{\sum_{i=1}^{i=3} (\Phi_i(x) - \Phi_i(y))^2}$$

(where $\Phi_i(x)$ is the ith component of the triple $\Phi(x)$), then

(a) $\forall x, y, z [y \text{ Bet}_\mathcal{A} xz \leftrightarrow d_\Phi(x, y) + d_\Phi(y, z) = d_\Phi(x, z)]$
(b) $\forall x, y, z, w [xy \text{ Cong}_\mathcal{A} zw \leftrightarrow d_\Phi(x, y) = d_\Phi(z, w)]$.

(U_E) Given any model of the axiom system and any two functions Φ and Φ' whose domain is the domain of the model: if Φ meets the conditions of the representation theorem (i.e. of (R_E)), then Φ' meets those conditions if and only if it has the form $T \circ \Phi$; where T

is a Euclidean transformation of $\mathbb{R}^3$, i.e. a transformation that can be obtained by some combination of shift of origin, reflection, rotation of axes, and multiplication of all coordinates by a positive constant (and where ∘ indicates functional composition).

We can state the second result more briefly by saying that the representation function guaranteed by (R_E) is *unique up to Euclidean transformation but no further*. This result gives an *explanation* of the fact that the laws of Euclidean geometry, when stated in terms of coordinates, are invariant under shift of origin, reflection, rotation, and multiplication of all distances by a constant factor: if we assume that the *genuine facts* about Euclidean space are just the facts about betweenness and congruence laid down in Hilbert's axioms, and that the function of coordinates is simply to facilitate the deduction of facts about betweenness and congruence and the relations definable in terms of them, then it *follows* that in an extrinsic formulation of the laws of geometry in terms of coordinates, the laws will be invariant up to Euclidean transformations and no further.

What we would like is to do for space-time what has been done for space. That is, we want to come up with a system of 'intrinsic' axioms, more or less analogous to Hilbert's but involving somewhat different concepts, and to come up with a representation theorem that explains the legitimacy of coordinatizing space-time and a uniqueness theorem that explains why in the coordinatized treatment of space-time the laws of Newtonian mechanics will be invariant under just the coordinate transformations that they are in fact invariant under. Anyone with the least familiarity with Newtonian mechanics knows what the relevant class of transformations is: it is the class of *generalized Galilean transformations*, that is, the class of transformations that can be obtained by some combination of: (a) shift or origin; (b) reflection; (c) rotation of spatial axes leaving the temporal axis fixed; (d) multiplication of all spatial coordinates by a positive constant; (e) multiplication of all temporal coordinates by a positive constant; and (f) change of rest frame according to the rule

$$t' = t$$
$$x' = x + ut$$
$$y' = y$$
$$z' = z$$

for some constant u. (Note the inclusion in this list of (d) and (e). The advantage I mentioned which my approach has over tensor approaches is that these are included among the transformations under which the theory is invariant.[28]) So our uniqueness theorem for space-time is going to have to say that the representation function Φ guaranteed by the representation theorem is unique up to generalized Galilean transformation and no further.

The key to developing a system of axioms from which one can prove the needed representation and uniqueness theorems is to build on the results of Szczerba and Tarski on affine geometry. Szczerba and Tarski laid down an axiom system somewhat like Hilbert's,[29] but invoking only the notion of betweenness. Their representation theorem for 3-dimensional space was just like (RE), but with clause (b) dropped since no notion of congruence was part of the system; to get the representation theorem for 4-dimensional space you simply replace $\mathbb{R}^3$ by $\mathbb{R}^4$ and use

$$\sqrt{\sum_{i=1}^{i=4}(\Phi_i(x) - \Phi_i(y))^2}$$

as your definition of $d_\Phi(x, y)$.[30] The uniqueness theorem for affine space is like (U_E), except that in place of Euclidean transformations is a more general kind of transformation called affine transformations. Affine

---

[28] Of course, (d) and (e) will be included in the covariance group of Newtonian mechanics on a tensor formulation, but that is irrelevant: so will lots of transformations that are clearly not symmetries, i.e. under which the laws in their usual formulations are not invariant. (For a good discussion of the conceptual distinction between symmetry and covariance, see Friedman 1981: ch. 3.) The reason why tensor approaches leave (d) and (e) out of the class of symmetries is that despite the fact that the main motivation of the tensor approach is to eliminate the use of arbitrarily chosen coordinate systems in formulating the laws of physics, it does not eliminate the use of arbitrarily chosen units of distance (or of arbitrarily chosen units for scalar magnitudes generally).

[29] In particular, the axiom system I am referring to is a second-order axiom system, but is interpretable nominalistically for the same reason that Hilbert's is: cf. Chapter 4. (Szczerba and Tarski also give a weaker first-order axiom system, but I am confining my attention here to the second-order system for the reason stated near the end of Chapter 4.)

[30] Those used to tensor formulations might find the appearance of a distance function in the context of affine space puzzling; after all, there is no uniquely defined metric in such a space! But as I remarked on the previous page, there is no uniquely defined metric in Euclidean geometry either, on an approach which (unlike the tensor approach) is fully invariant. Even though there is no uniquely defined metric in either affine or Euclidean geometry, it is legitimate to invoke a distance function in the representation theorem: for the representation theorem is simply a device for invoking mathematics extrinsically, to simplify calculations; and the distance function invoked in it is not invoked in the nominalistic theory that serves as our intrinsic explanation. (The right-hand side of the representation theorem for 'Bet' could also be formulated without invoking a distance function, but the formulation in terms of distance is easier to state.)

transformations preserve straight lines, and parallelism among lines, and the betweenness relation among points on a line, and the congruence relation among points on the same line (or more generally, the congruence or lack of congruence between x, y and z, w when x and y lie on a line parallel to a line containing z and w); they don't preserve perpendicularity, or congruence generally.

Generalized Galilean transformations are a special case of affine transformations. That they preserve straight lines, and betweenness on a line, and so forth, is obvious for purely spatial lines, i.e. lines all of whose points are simultaneous. How about for lines that are not purely spatial? It is easy to see that in the usual coordinatization, a non-spatial straight line is simply the path of an inertial coordinate system, and inertiality is preserved under generalized Galilean transformations. The betweenness relation on such a straight line has the obvious interpretation, and that relation too is preserved under generalized Galilean transformations. Two non-spatial straight lines are parallel if one is at rest relative to the other; and if x and y are on one line and z and w on a parallel line, then xy is congruent to zw if and only if the temporal separation between x and y is equal to that between z and w. These relations too are preserved by generalized Galilean transformations. So all generalized Galilean transformations are affine transformations;[31] the converse is not true, so what we want to do is to add a few primitive notions and axioms to the Szczerba-Tarski system, in such a way that we get representation and uniqueness theorems corresponding to the more restricted class of transformations.

The primitives we need, besides betweenness, are a two-place simultaneity relation and a four-place spatial congruence relation, with the property that xy S-Cong zw only if x is simultaneous to y and z is simultaneous to w.[32] Given that these are the primitives, it is clear what

---

[31] The preceding is a very redundant way to establish this: all we really need to check is that betweenness (which requires collinearity) is preserved by generalized Galilean transformations. But the redundancy is useful for giving an intuitive feeling for the significance of the 4-dimensional geometric claims.

[32] I have not introduced a temporal congruence predicate, because it is definable from betweenness and simultaneity (see (h) in this note). (The kind of spatial congruence relation considered here would also be definable from betweenness alone, if there were only one spatial dimension.) A good way to arrive at one's choice of primitives is to look for a set of primitives whose coordinate representations form a complete set of invariants for the class of generalized Galilean transformations: that is (a) each primitive must have a coordinate representation which is invariant under these transformations; and (b) for each transformation that isn't a generalized Galilean transformation, the coordinate representation of at least

our representation theorem is going to have to look like: it is going to have to be the 4-dimensional variant of (R_E), except with clause (b) dropped and the following two clauses added:

(c) $\forall x, y [x \text{ Simul } y \leftrightarrow \Phi_4(x) = \Phi_4(y)]$
(d) $\forall x, y, z, w [xy \text{ S-Cong } zw \leftrightarrow \Phi_4(x) = \Phi_4(y) \wedge \Phi_4(z) = \Phi_4(w) \wedge d_\Phi(x, y) = d_\Phi(z, w)]$.

And our uniqueness theorem is going to have to say that the representation function satisfying this representation theorem is unique up to

one of the primitives must fail to be invariant under that transformation. If these conditions were not met, the uniqueness theorem couldn't possibly hold.

For later reference I will give some definitions:

(a) Coll (x, y, z), meaning intuitively that x, y, and z lie on a line, is defined as

  y Bet $xz \vee x$ Bet $yz \vee z$ Bet xy.

(b) Coll (x, y, z, w) is defined as

  Coll $(x, y, z) \wedge$ Coll $(x, y, w) \wedge$ Coll $(x, z, w) \wedge$ Coll (y, z, w).

(c) Coplan (x, y, z, w), meaning intuitively that x, y, z, and w lie on a plane, is defined as

  $\exists u \{[\text{Coll } (u, x, y) \wedge \text{Coll } (u, z, w)] \vee [\text{Coll } (u, x, z) \wedge \text{Coll } (u, y, w)] \vee [\text{Coll } (u, x, w) \wedge \text{Coll } (u, y, z)]\}$.

(d) Cohyp (x, y, z, v, w), meaning intuitively that x, y, z, v, and w lie on a 3-dimensional hyper-plane, is defined as

  $\exists u \{[\text{Coplan } (u, x, y, z) \wedge \text{Coll } (u, v, w)] \vee [\text{Coplan } (u, x, y, v) \wedge \text{Coll } (u, z, w)] \vee [\text{Coplan } (u, x, y, w) \wedge \text{Coll } (u, z, v)]\}$.

(e) xy Par zw, meaning intuitively that $x \neq y$ and $z \neq w$ and the line passing through x and y is parallel to (or identical to) the line passing through z and w, is defined to be

  Coplan $(x, y, z, w) \wedge \neg \exists u [\text{Coll } (x, y, u) \wedge \text{Coll } (z, w, u)]$.

(f) Parallelogram (x, y, z, w), meaning intuitively that x, y, z, and w are vertices of a parallelogram with x opposite z, is defined as

  xy Par $zw \wedge xw$ Par yz.

(g) xy Par-Cong zw, meaning intuitively that x and y are on a line parallel to a line through z and w and the distance from x to y is equal to that from z to w, is defined as

  $(x = y \wedge z = w) \vee \exists u, v [\text{Parallelogram } (x, y, u, v) \wedge (\text{Parallelogram } (z, w, u, v) \vee \text{Parallelogram } (z, w, v, u))]$.

These are all affine-invariant notions, since they're defined from betweenness alone. Now using simultaneity as well, we can define temporal congruence:

(h) xy t-Cong zw is defined as

  $\exists x', y', z', w' [x \text{ Simul } x' \wedge y \text{ Simul } y' \wedge z \text{ Simul } z' \wedge w \text{ Simul } w' \wedge x'y' \text{ Par-Cong } z'w']$.

generalized Galilean transformations but no further. Given the Szczerba-Tarski axiom on 'Bet', it is quite trivial[33] to impose requirements on the two new primitives 'Simul' and 'S-Cong' so as to get the desired representation and uniqueness theorems.

The position that we arrive at, then, is that the only spatio-temporal relations needed to describe Newtonian space-time are the three invoked in this axiom system; all other genuine spatio-temporal relations are defined in terms of them, and relations which on a coordinate description of space-time might look genuine—e.g. being at absolute rest, or having a temporal separation of exactly 1, and so forth—really are not genuine but dependent on an arbitrary choice of coordinate system or distance function. By the representation theorem, the coordinate system and the distance function can be viewed as merely devices for deriving conclusions about spatio-temporal betweenness, simultaneity, and spatial congruence, conclusions which could be derived without ever bringing in numbers at all.

Of course, the conclusions arrived at so far are rather limited: they show the possibility of a nominalistic account of the structure of space-time, but they do not show that when we use space-time to develop broader theories (e.g. theories that describe the motion of particles by differential equations, theories that postulate scalar fields governed by other differential equations, and so forth), a nominalistic account of those broader theories is possible too. Nevertheless, I think that this more general conclusion is correct; and I will argue for it in the next two chapters, by building on the ideas developed so far.

---

[33] The simultaneity axioms are:
(a) $\exists a, b, c, d [a \text{ Simul } b \wedge b \text{ Simul } c \wedge c \text{ Simul } d \wedge \neg\text{Coplan}(a, b, c, d)]$.
(b) $\forall a, b, c, d [a \text{ Simul } b \wedge b \text{ Simul } c \wedge c \text{ Simul } d \wedge \neg\text{Coplan}(a, b, c, d) \rightarrow \forall x, y [x \text{ Simul } y \leftrightarrow x = y \vee \exists e (e \neq a \wedge xy \text{ Par } ae \wedge \text{Cohyp}(a, b, c, d, e))]]$.

(Cf. the previous footnote for the definitions of 'Coplan' and 'Cohyp'.) To axiomatize S-Congruence, start with the usual congruence axioms that when added to the axioms of 3-dimensional affine geometry give you a complete axiomatization of 3-dimensional Euclidean geometry, and in each of these axioms restrict all variables so that all points mentioned are required to be simultaneous with each other. These axioms, plus also

(a) $xy \text{ S-Cong } zw \rightarrow x \text{ Simul } y \wedge z \text{ Simul } w$
(b) $xy \text{ S-Cong } zw \leftrightarrow \exists z', w' (xy \text{ S-Cong } z'w' \wedge x \text{ Simul } z' \wedge z'w' \text{ Par-Cong } zw)$,

where Par-congruence is as defined in the previous footnote, suffice for the desired representation and uniqueness theorems. Doubtless there are more elegant axiomatizations of 'S-Cong' than this, but this one has the advantage of being *obviously* adequate.

7
A Nominalistic Treatment of Quantities, and a Preview of a Nominalistic Treatment of the Laws Involving Them

I have described a nominalistic treatment of space-time; next we have to deal with entities that exist within space-time, e.g. various scalar fields such as temperature or gravitational potential.

A possible approach to a coordinate-independent treatment of, say, temperature, would be to introduce a continuum of temperature properties, each one the property of having such and such specific temperature. One could then describe the structure of that system of properties not via numbers, but via certain intrinsic relations among them, say the relations of betweenness and congruence; and one could impose axioms on these notions to guarantee that there was a 1–1 function mapping the temperature properties into the reals, and that such a function was unique up to linear transformation. There is a certain conception of properties (that of Putnam 1970) on which this approach would be at least arguably a nominalistic one; but I prefer a different strategy, which doesn't invoke temperature properties but which makes do with space-time points (or more generally, space-time regions) as the only entities.

My approach is not to introduce betweenness and congruence relations *among temperature properties*, but to introduce temperature-betweenness and temperature-congruence relations *among space-time points*. That is, we will have a three-place relation Temp-Bet, with y Temp-Bet xz

meaning intuitively that y is a space-time point at which the temperature is (inclusively) between the temperatures of points x and z; and a four-place relation Temp-Cong, with xy Temp-Cong zw meaning intuitively that the temperature difference between points x and y is equal in absolute value to the temperature difference between points z and w. Also, if we are going to want to formulate laws which in their extrinsic formulations are not invariant under temperature reversal (i.e. under a systematic replacement of low temperatures by high temperatures and conversely), we will need a two-place predicate Temp-Less, where x Temp-Less y means that the point x is lower than or equal in temperature to point y. (When Temp-Less is used, Temp-Bet of course becomes definable; but in order to make most of the formal development independent of whether the laws are invariant under reversal, it is convenient in the exposition that follows to keep Temp-Bet as a primitive in either case.)

We now want to impose axioms on these relations, which will give us representation and uniqueness theorems more or less analogous to the Hilbert theorems (R_E) and (U_E) of the previous chapter. The theorems we want are

(R_{Temp}) A structure $\langle \mathcal{A}, \text{Temp-Bet}_\mathcal{A}, \text{Temp-Cong}_\mathcal{A} \rangle$ or $\langle \mathcal{A}, \text{Temp-Bet}_\mathcal{A}, \text{Temp-Cong}_\mathcal{A}, \text{Temp-Less}_\mathcal{A} \rangle$ is a model of the axioms if and only if there is a function Ψ from $\mathcal{A}$ onto an interval (connected subset with more than one element) of real numbers, such that

(a) $\forall x, y, z[y \text{ Temp-Bet}_\mathcal{A} xz \leftrightarrow$ either $\Psi(x) \leq \Psi(y) \leq \Psi(z)$ or $\Psi(z) \leq \Psi(y) \leq \Psi(x)]$
(b) $\forall x, y, z, w[xy \text{ Temp-Cong}_\mathcal{A} zw \leftrightarrow |\Psi(x) - \Psi(y)| = |\Psi(z) - \Psi(w)|]$,

and if Temp-Less is used as a primitive,

(c) $\forall x, y[x \text{ Temp-Less}_\mathcal{A} \leftrightarrow \Psi(x) \leq \Psi(y)]$.

(U_{Temp}) Given any model of the axiom system and any two functions Ψ and Ψ' whose domain is the domain of the model: if Ψ meets the conditions of (R_{Temp}), then Ψ' meets those conditions if and only if it has the form $P \circ \Psi$; where

if Temp-Less is not used as a primitive, P is a linear transformation of the reals, i.e. a function of form $ax + b$ where b is a real and a is a non-zero real; and if Temp-Less is used as a primitive, P is a *positive* linear transformation, i.e. a linear transformation in which the constant a is greater than zero.

Note that unlike the case of Euclidean geometry or the geometry of space-time, we don't demand that the representation functions be 1-1, for the obvious reason that different space-time points may have the same temperature. Also, we don't demand that the representation function be the entire set of reals. We do demand that it be a connected set of reals, for a simple reason: temperature and other scalar fields used in physics are assumed to be continuous, and this guarantees that if point x has temperature $\Psi(x)$ and point z has temperature $\Psi(z)$ and r is a real number between $\Psi(x)$ and $\Psi(z)$, then there will be a point y spatio-temporally between x and z such that $\Psi(y) = r$. I have also demanded that the range of Ψ contain more than one real number, simply because doing so avoids the need to worry about tedious special cases in stating definitions later on, and because the case of a scalar with only one value is of no interest.

The task of getting an axiom system for Temp-Bet and Temp-Cong and perhaps Temp-Less that will give rise to the desired representation and uniqueness theorems is a problem that has in essence been solved by others. There is in fact a large body of known results about how to axiomatize something so as to get desired representation and uniqueness theorems: these results form the major part of what is known as 'the theory of measurement'. This name reflects a philosophical bias quite different from mine: it reflects a concern with something like operational definitions, rather than with axiomatizing science without the use of numbers. But 'measurement theory' has progressed largely by ignoring characteristic features of measurement (such as measurement errors) and focussing on such questions as: what must the intrinsic facts about temperature differences between physical objects be if it is appropriate to think of temperature as being represented by real numbers? And except for the fact that I am substituting space-time points for physical objects, this is in effect the question I am now asking.

An excellent survey of the kind of results now available in modern measurement theory is given in Krantz, Luce, Suppes, and Tversky 1971. Many of the topics in that book are of some relevance to the project of nominalizing physics; of immediate interest is the treatment they give of 'absolute difference structures' in their section 4.10. For their system is, in effect, an axiom system involving betweenness and congruence[34] which gives something very close to the representation and uniqueness theorems for scalar fields like temperature that were required above. It is quite easy[35] to modify their system so as to give the representation and uniqueness theorems we want, when Temp-Less is not used as a primitive. And when Temp-Less is used as an additional primitive, it is easy[36] to add to the Krantz axioms some new axioms relating this to the other primitives so as to again get the representation and uniqueness theorem we want; in fact an earlier section of the Krantz book, on 'algebraic difference structures', shows in effect how this is to be done. For future

[34] Actually they use a single primitive, $xy \leq zw$, meaning that the absolute value of the scalar difference between x and y is less than or equal to the absolute value of the scalar difference between z and w. But their system is convertible to one using betweenness and congruence as primitives; or equivalently, my remarks could be easily modified so as to make use of their primitive instead of betweenness and congruence.

[35] Two modifications of the Krantz axiomatizations are required. In the first place, the Krantz system yields a representation theorem in which the range of the scalar function is not required to be connected. But this is easy to fix: add an axiom saying that for any two points x and z there is a third point y such that y Temp-Bet xz and xy Temp-Cong yz; and replace the Archimedean axiom by a Dedekind continuity axiom. Since the Krantz system already contains an axiom that allows subtraction (i.e. with the consequence that if r_1, r_2, r_3, and r_4 are in the range of the scalar and $|r_3 - r_4| < |r_1 - r_2|$, then there is an r in the range of the scalar between r_1 and r_2 such that $|r_1 - r| = |r_3 - r_4|$), these modifications clearly suffice for the range of the scalar to be connected.

The second modification that is required is due to the fact that their system leads to a representation function which is 1-1. However, it is easy to modify the system so that this is not a consequence: instead of supposing that x Bet yy and $\exists z(xy$ Cong $zz)$ each imply $x = y$, require only that x Bet $yy \leftrightarrow \exists z(xy$ Cong $zz)$, and that x Bet yy is an equivalence relation, and that substitutivity of equivalents never affects betweenness or congruence. This axiomatization (which is actually a bit redundant) obviously works, for the equivalence classes satisfy the original axiom system, and a 1-1 representation function whose domain is the set of equivalence classes induces a not-necessarily 1-1 representation function on the space-time points themselves.

Incidentally, it may alleviate confusion to point out that my style of stating representation theorems is different from that of Krantz et al. My representation theorems say that a structure (of the appropriate type) is a model of such and such a theory if and only if there is a representation function of such and such a sort; theirs say *only if* rather than *if and only if*, and the statements of their theorems would be false if you replaced 'only if' by 'if and only if' because of their use of what they call 'non-necessary axioms'. The reason I have been able to avoid 'non-necessary axioms', and hence make 'if and only if' statements, is that I have strengthened the system so as to require that the ranges of scalar functions be connected. One of the virtues of the space-time approach to these matters is that it allows that.

[36] Simply add that Temp-Less is transitive and connected, and that

y Temp-Bet $xz \leftrightarrow (x$ Temp-Less $y \wedge y$ Temp-Less $z) \vee (z$ Temp-Less $y \wedge y$ Temp-Less $x)$.

reference it will be convenient to have names for the axiom systems that arise by modifying the Krantz axioms in these ways: I'll call them the axiom systems for unordered scalar fields and for ordered scalar fields.

Of course, nothing in all this turns on the scalar in question being temperature, rather than gravitational potential or some other scalar. In the future then, I will write 'Scal-Bet', 'Scal-Cong', etc. for the predicates, to emphasize that we're dealing with an arbitrary scalar. (In an actual physical theory one might of course have more than one scalar; so there will be different families of betweenness and congruence predicates, one for each such scalar. This could be indicated in the notation by a subscript on the prefix 'Scal', but I won't bother to do so.) In order to distinguish our spatio-temporal betweenness predicate from the scalar-betweenness predicate(s), I'll write the former as st-Bet.

If we now introduce a joint axiom system JAS that includes the spatio-temporal primitives and the temperature (or other scalar) primitives together, both defined on the same domain (which we think of as the set of space-time points), and if we impose the appropriate axioms for each, then for any model of the combined system there is both a 1-1 spatio-temporal function Φ onto $\mathbb{R}^4$ and a scalar-representation function Ψ onto an interval, each function unique up to (but only up to) the appropriate class of transformations. Now, physical laws governing a scalar like temperature or gravitational potential are often expressed as laws about a scalar function T mapping quadruples of real numbers into real numbers (the quadruples of reals in the domain representing space-time location and the numbers in the range representing temperatures or gravitational potentials or whatever). It should be clear that such a function T is precisely $\Psi \circ \Phi^{-1}$ (see Figure 2). This suggests that *laws about T* (e.g. that it obeys such and such a differential equation) *could be restated as laws about the interrelation of Φ and Ψ. And since Φ and Ψ are generated by the basic predicates* Scal-Bet, Scal-Cong, st-Bet, Simul, S-Cong, and perhaps Scal-Less, *it is natural to suppose that the laws about T could be further restated in terms of these latter predicates alone.*

Of course, we can't hope to express all properties of T in terms of these five (or six) predicates: only those features of T that are invariant under both generalized Galilean transformations of the spatio-temporal coordinates and under linear (or positive linear) transformations of the temperature scale could ever be so expressed. But that's alright, for it is only such invariant properties of T that are of any physical importance

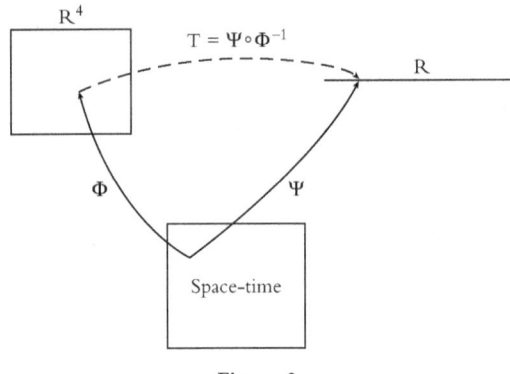

Figure 2

anyway. What we must hope, then, is that given some law (e.g. a differential equation) involving T, we can find some nominalistic formulation involving our five (or six) basic predicates that gives the full invariant content of the law.[37]

In the next chapter I will show that in many cases—and, I suspect, in all—it is indeed possible to do this. This will be the key to nominalizing Newton's theory of gravitation.

---

[37] Whether or not we use the sixth predicate depends on how invariant the law is taken to be. Physical laws involving the gravitational potential are not in general invariant under reflection, so we will need to invoke a predicate 'Grav-Less' in formulating physics nominalistically; but in sketching the approach it is better to leave open the question of whether there is a 'less than' primitive, so as to make the approach also apply to scalar fields that enter only into laws that are invariant under reversal.

8
Newtonian Gravitational Theory Nominalized

A. Continuity

I will begin the illustration of the ideas of the paragraph before last by a simpler example than a differential equation. Suppose we want to say nominalistically that the scalar function T is a continuous function. In 'saying this nominalistically', we are not allowed to talk about T at all: T, after all, is a function, and hence not a nominalistically admissible entity. The sentence CONT which we will use to 'say that T is continuous' will in fact quantify only over space-time points and space-time regions, and will use only the basic predicates listed in the third from last paragraph of the preceding chapter. We will prove

(1) For any model of the joint axiom system JAS for space-time and the scalar quantity in question, and any representation functions Φ and Ψ (for space-time and the scalar quantity, respectively), the new claim CONT is true in the model if and only if T (i.e. $\Psi \circ \Phi^{-1}$) is a continuous function.

This in effect will extend our representation theorem to the larger axiom system that includes the new continuity claim CONT. The uniqueness theorem will extend automatically, since the new continuity axiom will involve only the basic predicates used in the original axiom system.

Such a nominalistic continuity axiom is easy enough to find. (The following formulation of it quantifies over space-time regions. A formulation quantifying over space-time points only is possible too, but I will not bother to give it since regions appear to be needed later on anyway, and since as already remarked there doesn't seem to be much point in denying the existence of regions while admitting the existence of points.)

First a preliminary definition: define $x \approx_{Scal} y$ as x Scal-Bet yy; intuitively it means that x and y have the same temperature, or the same gravitational potential, or the same value of whatever scalar quantity is in question. Now call a region R *scalar-basic* (or temperature-basic or gravitational-potential-basic, when we want to distinguish the scalar in question from other scalars that will occur in the theory) if and only if there are distinct points x and y such that either

(a) R contains precisely those points z such that z Scal-Bet xy and not $(z \approx_{Scal} x)$ and not $(z \approx_{Scal} y)$ (cf. Figure 3(a)); or
(b) R contains precisely those points z such that y Scal-Bet xz and not $(z \approx_{Scal} y)$ (cf. Figure 3(b)).

Figure 3(a)   Figure 3(b)

In terms of a representation function Ψ for temperature, then, R is temperature-basic if and only if either it consists of all points whose temperature is exclusively between $\Psi(x)$ and $\Psi(y)$, or it consists of all points whose temperature is greater than $\Psi(y)$, or it consists of all points whose temperature is less than $\Psi(y)$. (So the temperature-basic regions will correspond to the sets of space-time points that are Ψ^{-1}-images of basic open sets in the usual topology of the interval that is the range of Ψ.[38] This correspondence will hold for any such representation function Ψ.)

We can similarly, using the notion of st-Bet, characterize the regions of space-time that are *spatio-temporally basic*, i.e. that are mapped onto basic open sets of $\mathbb{R}^4$ (say, interiors of tetrahedrons in $\mathbb{R}^4$) by any spatio-temporal representation function Φ.

We then 'say that T (i.e. $\Psi \circ \Phi^{-1}$) is continuous at $\Phi(x)$' by saying that for any temperature-basic region that contains x, there is a spatio-temporally basic subregion that contains x; and we 'say that T

---

[38] Temperature-basic sets may be mapped by Ψ onto semi-closed sets of real numbers, but only if the included end-point is the largest or smallest temperature value attained anywhere throughout the whole space.

is continuous' by saying that this holds for every space-time point x.[39] This claim CONT 'says that T is continuous' in the sense that for the claim, (1) holds.

But in 'saying that the scalar field is continuous' we haven't mentioned any specific T (or any specific Ψ or Φ): doing that would not only violate nominalism, it would also involve a particular choice of coordinate system for space-time and a particular choice of scale for temperature. Nor have we 'quantified over arbitrary choices', i.e. talked about all functions T that would result by making arbitrary choices of coordinate system and of scale in one or another way. Rather, we have specified the continuity of temperature with respect to space-time in a completely intrinsic way, a way that never mentions spatio-temporal coordinates or temperature scales. In my view this fully intrinsic character of the method makes it very attractive even independently of nominalistic scruples.

The final thing to note about the treatment of the continuity of T is that we made use only of the affine properties of space-time, i.e. of the properties that depend only on the betweenness relation and not on the simultaneity or spatial congruence relations. In most of the mathematical development to follow—e.g. in the nominalistic treatment of differentiation—the same will be true. This is important, for it will mean that these developments can be carried over without any change at all to physical theories that do not postulate a Newtonian space-time but postulate some other space-time with a (flat, globally $\mathbb{R}^4$) affine structure instead; e.g., these developments will go over without alteration to the special theory of relativity. (Some of the other mathematical developments which do involve aspects of Newtonian space-time other than its affine properties—e.g. the treatment of gradients and Laplacians—will also go over to special relativity with very little change; and I *believe* that without too much trouble all the mathematical developments to follow could be generalized to space-time with a more general sort of affine structure than considered here (i.e. space-times which don't obey all the Szczerba-Tarski axioms and indeed which require a more complicated set of primitives), such as the space-time of general relativity. See the last paragraph of note 44.) Analogously, the above treatment of continuity

---

[39] Observe that the basic idea of this approach to continuity is to use two topologies on the same set (the set of space-time points), rather than topologies on two different sets that are related by a function. That is the secret of how quantifying over functions is avoided.

does not rely on the primitive 'Scal-Less', and again we will do without that predicate in our mathematical development, for the sake of generality, whenever possible.[40]

B. Products and Ratios

Before dealing with differentiation proper, we must deal with the comparison of products or ratios. For instance, suppose we want to say nominalistically that the result of multiplying a certain pair of intervals is less than the result of multiplying a certain other pair of intervals. Obviously there is no hope of saying this nominalistically unless certain conditions on units are met. For instance, there is no hope of saying nominalistically that the result of multiplying two spatio-temporal intervals is less than the result of multiplying two temperature intervals, for the truth of such a statement would depend on the choice of a temperature scale and of spatio-temporal coordinates, and in a nominalistic treatment (of the sort being proposed here) only invariant statements are possible. What we can hope to say is that the result of multiplying one spatio-temporal interval with one temperature interval is less than the result of multiplying another spatio-temporal interval with another temperature interval. At least, we can hope to say this if the two spatio-temporal intervals are themselves objectively comparable; and they will be (in any affine space) when both lie on the same straight line.[41] What we want then is a nominalistic statement with eight free variables, which we may abbreviate suggestively as

(2) $\quad |x_1 x_2 y_1 y_2| <_{st,Scal} |u_1 u_2 v_1 v_2|$,

such that for any choice of a spatio-temporal representation function Φ and a scalar representation function Ψ, we can prove the biconditional

---

[40] Although in the treatment of continuity we also avoided Scal-Cong, it is not in general possible to do much without Scal-Cong. Scal-Cong, unlike spatial congruence, is an affine notion: that is, although in two or more dimensions the affine properties of space (which include the Par-congruence relation of note 32) are definable in terms of betweenness, this isn't true in one dimension; and in one dimension there is no distinction between the affine notion of Par-congruence and the most general congruence relation. So for a one-dimensional structure like temperature, congruence is an affine notion, and there is no way to avoid it in developing calculus.

[41] Or on parallel lines; but for simplicity I confine myself to intervals on the same line.

(3) $|x_1x_2y_1y_2| <_{st,Scal} |u_1u_2v_1v_2|$ if and only if x_1, x_2, u_1, and u_2 lie on a single line and $d_\Phi(x_1,x_2) \cdot |\Psi(y_1) - \Psi(y_2)| < d_\Phi(u_1,u_2) \cdot |\Psi(v_1) - \Psi(v_2)|$.

Here d_Φ, is the spatio-temporal distance function

$$\sqrt{\sum_{i=1}^{i=4}(\Phi_i(x) - \Phi_i(y))^2}.$$

Note that although d_Φ is highly non-invariant under affine transformations of the spatio-temporal coordinate system, the right-hand side of (3) in which it appears is invariant under affine transformations; so this is one of those cases where we ought to expect that we won't need non-affine-invariant notions in the definition of (2) (i.e. we ought to expect that we not only won't need to use 'd_Φ' in the definition of (2), we won't need 'Simul' or 'S-Cong' either). Similarly, since the right-hand side of (3) is invariant under all linear transformations of the Ψ-scale (not just the positive linear transformations), this is one of those cases where we should hope to do without use of the predicate 'Scal-Less' in the definition. So we want to be able to give a nominalistic definition of (2) that meets these additional constraints, and such that the biconditional (3) is provable.

This task is easily carried out, if we allow ourselves sufficient logical machinery. First we define a *spatio-temporally equally spaced region* (see Figure 4) as a region R all of whose points lie on a single line, and such that for every point x of R which lies strictly st-between two points of R, there are points y and z of R such that

(a) exactly one point of R is strictly st-between y and z, and this point is x; and
(b) xy Par-Cong xz.

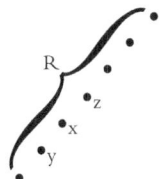

Figure 4

(Par-congruence, i.e. congruence along parallel lines, is defined in note 32.) My definition allows equally spaced regions to be infinite as well as finite; however, it is really only the finite ones we need.

The notion of a region *equally spaced in temperature* is analogous, except that st-betweenness is replaced throughout by temperature-betweenness and Par-congruence by temperature-congruence. (The requirement that all points of the region lie on a single line can be dropped; for the claim that results from this when temperature-betweenness is substituted for st-betweenness in the definition of lying on a single line is vacuous, since the temperature-ordering is 1-dimensional.) Given this, we can define (2) as follows (see Figure 5):

(4) $u_1 \neq u_2$ and not ($v_1 \approx_{Scal} v_2$); and

if $x_1 \neq x_2$ and not ($y_1 \approx_{Scal} y_2$), then there are R_{st} and R_{Scal} such that

  (i) R_{st} is an st-equally-spaced region containing x_1 and x_2;
  (ii) R_{Scal} is a scalar-equally-spaced region containing y_1 and y_2;
  (iii) there are a, b in R_{st} such that u_1 and u_2 are st-between a and b, and there are c, d in R_{Scal} such that v_1 and v_2 are Scal-between c and d;
  (iv) there are just as many points of R_{st} that are st-between x_1 and x_2 as there are points of R_{Scal} that are Scal-between v_1 and v_2; and
  (v) there are fewer points of R_{Scal} that are Scal-between y_1 and y_2 than there are points of R_{st} that are st-between u_1 and u_2.

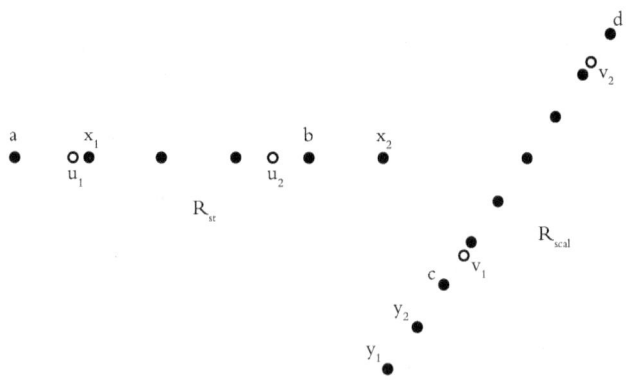

Figure 5

All the notions in this definition are affine-invariant, and if we define (2) by (4) then it is routine[42] to prove the required representation theorem (3). (The question may be raised whether our definition (4) is genuinely nominalistic, due to the cardinality comparisons that occur in it; I defer this issue until Chapter 9.)

We also need to define a relation $|x_1 x_2 y_1 y_2| >_{st,scal} |u_1 u_2 v_1 v_2|$ with a representation theorem like (3) but with '<' replaced by '>', but obviously this can be done analogously. And given '$<_{st,Scal}$' and '$>_{st,Scal}$', '$=_{st,Scal}$' can be defined in terms of them in the obvious way.

Finally, I note for later use that the definition of one product being less than (or greater than, etc.) another can be straightforwardly generalized to the case where the products are of more than two factors. For instance, we can define by means quite analogous to (4) a formula

---

[42] We can assume that $x_1, x_2, u_1,$ and u_2 all lie on a single line, and that $x_1 \neq x_2, u_1 \neq u_2$, not $y_1 \approx_{Scal} y_2$, and not $v_1 \approx_{Scal} v_2$; for when these conditions *aren't* met, it is *clear* that biconditional (3) holds. Let r_x and r_u be $d_\Phi(x_1, x_2)$ and $d_\Phi(u_1, u_2)$ respectively, and let r_y and r_v be $|\Psi(y_1) - \Psi(y_2)|$ and $|\Psi(v_1) - \Psi(v_2)|$ respectively. Let R_{st} and R_{Scal} be regions meeting all the conditions in (4) except possibly the last. Let N_x be the number of points in R_{st} (inclusively) between x_1 and x_2, and define $N_u, N_y,$ and N_v similarly (using R_{Scal} in place of R_{st} for N_y and N_v); since we're assuming that $x_1 \neq x_2$ and not $y_1 \approx_{Scal} y_2$, N_x and N_y are at least 2. Finally, let μ_Φ be the Φ-distance between adjacent points of R_{st}, and μ_Ψ be the absolute difference between the Ψ-values of Scal-adjacent points in R_{Scal}. Then

$(N_x - 1)\mu_\Phi = r_x$
$(N_y - 1)\mu_\Psi = r_y$
$(N_u - 1)\mu_\Phi \leq r_u < (N_u + 1)\mu_\Phi$
$(N_v - 1)\mu_\Psi \leq r_v < (N_v + 1)\mu_\Psi$

Also $N_x = N_y$. We need: (a) that if $N_y < N_u$, then $r_x r_y < r_u r_v$; and conversely (b) that if $r_x r_y < r_u r_v$ then for some choice of R_{st} and R_{Scal} meeting the conditions, $N_y < N_u$.

To prove these, note that the indented inequalities give

$$\frac{(N_u - 1)(N_v - 1)}{(N_x - 1)(N_y - 1)} \leq \frac{r_u r_v}{r_x r_y} < \frac{(N_u + 1)(N_v + 1)}{(N_x - 1)(N_y - 1)}$$

which together with $N_x = N_y$ gives

$$\frac{N_u - 1}{N_y - 1} \leq \frac{r_u r_v}{r_x r_y} < \left(\frac{N_u - 1}{N_y - 1} + \frac{2}{N_y - 1}\right)\left(1 + \frac{2}{N_x - 1}\right).$$

The left-most inequality here establishes (a). For (b), note that as we decrease the spacing between the points of R_{st}, we increase N_x; and then to keep $N_x = N_y$ we must also decrease the spacing between points of R_{Scal}, and so N_y must increase too. We see that by decreasing the spacing between the points of R_{st} and holding $N_x = N_y$ fixed, we can get the right-hand side of the inequality arbitrarily close to $\frac{N_u-1}{N_y-1}$. So if $\frac{r_u r_v}{r_x r_y} > 1$, then we can see that for sufficiently fine-meshed R_{st} and R_{Scal}, $\frac{N_u-1}{N_y-1} > 1$ also; so $N_u > N_y$, establishing (b).

(2′) $|x_1 x_2 y_1 y_2 z_1 z_2| <_{st,st,Scal} |u_1 u_2 v_1 v_2 w_1 w_2|$

for which we can prove that for any representation functions Φ and Ψ for space-time and our scalar respectively, the bi-conditional

(3′) $|x_1 x_2 y_1 y_2 z_1 z_2| <_{st,st,Scal} |u_1 u_2 v_1 v_2 w_1 w_2|$ if and only if:
x_1, x_2, u_1, and u_2 all lie on a line, and y_1, y_2, v_1, and v_2 all lie on (the same or another) line, and
$d_\Phi(x_1, x_2) \cdot d_\Phi(y_1, y_2) \cdot |\Psi(z_1) - \Psi(z_2)| \quad < d_\Phi(u_1, u_2) \cdot d_\Phi(v_1, v_2) \cdot |\Psi(w_1) - \Psi(w_2)|$.[43]

C. Signed Products and Ratios

So far we've been talking of products of *absolute values*, but a more general kind of product comparison is also useful (even when we are dealing with unordered scalar fields). Platonistically, these new product comparisons are most easily made if we introduce a new kind of representation function. Suppose we are talking about points on a single line L. Our old coordinatization Φ of space assigns points of $\mathbb{R}^4$ to points of L; let's introduce a new coordinatization Φ_L that assigns real numbers to points of L, and that is 'compatible with' the old one Φ in the sense that for any points x and y on L, $|\Phi_L(x) - \Phi_L(y)| = d_\Phi(x, y)$. The choice of such a Φ_L is to some extent arbitrary: it is arbitrary which point of L is assigned 0, and which direction along L is the direction of increasing Φ_L values. So our only interest in relations expressed using Φ_L is in those relations that do not depend on these arbitrary choices (in addition to not depending

---

[43] The generalization of (4) is:

$u_1 \neq u_2 \wedge v_1 \neq v_2 \wedge \text{not}(w_1 \approx_{Scal} w_2)$, and
if $x_1 \neq x_2 \wedge y_1 \neq y_2 \wedge \text{not}(z_1 \approx_{Scal} z_2)$, then there are R_{st}, R'_{st} and R_{Scal} such that:

(i) R_{st} and R'_{st} are st-equally-spaced regions and R_{Scal} is a scalar-equally-spaced region;
(ii) x_1 and x_2 are in R_{st}, y_1 and y_2 are in R'_{st}, and z_1 and z_2 are in R_{Scal};
(iii) there are a, b in R_{st} such that u_1 and u_2 are st-between a and b, and there are c, d in R'_{st} such that v_1 and v_2 are st-between c and d, and there are e, f in R_{Scal} such that w_1 and w_2 are Scal-between e and f;
(iv) there are just as many points of R_{st} between x_1 and x_2 as of R'_{st} between v_1 and v_2;
(v) there are just as many points of R'_{st} between y_1 and y_2 as of R_{Scal} between w_1 and w_2; and
(vi) there are fewer points of R_{Scal} between z_1 and z_2 than of R_{st} between u_1 and u_2.

on the arbitrary choice of a particular Φ or Ψ). One such relation is the following:

(5) $(\Phi_L(x_2) - \Phi_L(x_1)) \cdot (\Psi(y_2) - \Psi(y_1))$ is exclusively between $(\Phi_L(s_2) - \Phi_L(s_1)) \cdot (\Psi(t_2) - \Psi(t_1))$ and $(\Phi_L(u_2) - \Phi_L(u_1)) \cdot (\Psi(v_2) - \Psi(v_1))$.

Here we are expressing not a comparison of products of absolute values of spatio-temporal intervals and scalar intervals, but a comparison of signed products of oriented spatio-temporal intervals and oriented scalar intervals. It is however a comparison that is invariant under choice of the orientations of Ψ and of Φ_L. What we now have to do is express this comparison nominalistically and without ever introducing arbitrary orientations.

More precisely, then, what we want to do is to nominalistically define a predicate

(6) $(x_1 x_2 y_1 y_2)$ E-Bet$_{st, Scal}$ $(s_1 s_2 t_1 t_2)(u_1 u_2 v_1 v_2)$

for which we can prove a representation theorem that says that (6) holds if and only if the points x_1, x_2, s_1, s_2, u_1 and u_2 all lie on a single line L, and for any representation functions Φ and Ψ and any coordinatization Φ_L of L compatible with Φ, (5) holds.

To define (6) it is useful to first define two other notions: first, $x_1 x_2$ Pos-Par $u_1 u_2$, meaning intuitively that the line-segment from x_1 to x_2 is parallel to and pointing in the same direction as the line-segment from u_1 to u_2 (with $x_1 \neq x_2$ and $u_1 \neq u_2$); second, $y_1 y_2$ Pos-Orient $v_1 v_2$, meaning intuitively that $\Psi(y_2) - \Psi(y_1)$ and $\Psi(v_2) - \Psi(v_1)$ are non-zero and have the same sign. Both these relations are easy to define nominalistically (and we don't need 'Scal-Less' to define the latter). Given these relations, one can define an eight-place relation $x_1 x_2 y_1 y_2$ Same-Sign $u_1 u_2 v_1 v_2$ as:

Either $x_1 x_2$ Pos-Par $u_1 u_2$ and $y_1 y_2$ Pos-Orient $v_1 v_2$, or $x_1 x_2$ Pos-Par $u_2 u_1$ and $y_1 y_2$ Pos-Orient $v_2 v_1$.

In the case when x_1, x_2, u_1 and u_2 lie on a single line L, this relation will hold if and only if $(\Phi_L(x_2) - \Phi_L(x_1))(\Psi(y_2) - \Psi(y_1))$ and $(\Phi_L(u_2) - \Phi_L(u_1))(\Psi(v_2) - \Psi(v_1))$ are non-zero and have the same sign. We also define $x_1 x_2 y_1 y_2$ Opp-Sign $u_1 u_2 v_1 v_2$ as $x_1 x_2 y_1 y_2$ Same-Sign $u_2 u_1 v_1 v_2$. One can now define (6) in terms of the simpler kind of product comparison in the preceding section, as follows:

x_1, x_2, u_1, u_2, s_1, and s_2 all lie on a single line;
and if $x_1 = x_2$ or $y_1 \approx_{Scal} y_2$ then $s_1s_2t_1t_2$ Opp-Sign $u_1u_2v_1v_2$;
and if $x_1 \neq x_2$ and not($y_1 \approx_{Scal} y_2$) then one of the following three conditions holds:

(a) $s_1s_2t_1t_2$ Same-Sign $x_1x_2y_1y_2$, and $u_1u_2v_1v_2$ Same-Sign $x_1x_2y_1y_2$, and either $|s_1s_2t_1t_2| <_{st,Scal} |x_1x_2y_1y_2| <_{st,Scal} |u_1u_2v_1v_2|$ or $|u_1u_2v_1v_2| <_{st,Scal} |x_1x_2y_1y_2| <_{st,Scal} |s_1s_2t_1t_2|$;

(b) $s_1s_2t_1t_2$ Same-Sign $x_1x_2y_1y_2$, and not ($u_1u_2v_1v_2$ Same-Sign $x_1x_2y_1y_2$), and $|x_1x_2y_1y_2| <_{st,Scal} |s_1s_2t_1t_2|$;

(c) not ($s_1s_2t_1t_2$ Same-Sign $x_1x_2y_1y_2$), and $u_1u_2v_1v_2$ Same-Sign $x_1x_2y_1y_2$, and $|x_1x_2y_1y_2| <_{st,Scal} |u_1u_2v_1v_2|$.

The desired representation theorem connecting (6) so defined with (5) is now easily provable.

(6) shows how to define one signed product being between two others. One can similarly define what it is for one signed product to equal another signed product. Finally, one can extend this result to signed products of three or more factors: e.g. in analogy to (2′) and (3′) at the very end of the previous section, we can define a formula

$$(x_1x_2y_1y_2z_1z_2) =_{st,st,Scal} (u_1u_2v_1v_2w_1w_2)$$

which holds if and only if x_1, x_2, u_1, and u_2 are all on a line L, and y_1, y_2, v_1, and v_2 are all on a line L', and

$$(\Phi_L(x_2) - \Phi_L(x_1))(\Phi_{L'}(y_2)-\Phi_{L'}(y_1))(\Psi(z_2)-\Psi(z_1)) = (\Phi_L(u_2) - \Phi_L(u_1))(\Phi_{L'}(v_2) - \Phi_{L'}(v_1))(\Psi(w_2) - \Psi(w_1)).$$

D. Derivatives

Now that such product comparisons are at hand, we can deal with the differentiability properties of our scalar function T ($= \Psi \circ \Phi^{-1}$). Suppose for instance that we want to say something about the existence of the partial derivatives of T at a given point, and the values of these partial derivatives there. We can't actually say that the partial derivatives have certain values, for this is not an invariant statement: it depends on the directions of the spatial and temporal axes, and the scale units for space, time, and temperature. So the first step is to find a way of stating the invariant content of the claim that the partial derivatives have such and such values. The secret is to ask not about the *values* of the *partial* derivatives, but for *comparisons* of the *directional* derivatives with the temperature intervals. That is, the statement

(7) the directional derivative of $T(=\Psi \circ \Phi^{-1})$ with respect to the vector $\Phi(a_2) - \Phi(a_1)$ exists at $\Phi(x)$ and has a value there equal to $\Psi(b_2) - \Psi(b_1)$

is invariant under generalized Galilean (in fact, affine) transformations of Φ and under linear transformations of Ψ, so it is suitable as the right-hand side of a representation theorem; let us then try to 'say it nominalistically', i.e. find a statement that can be the left-hand side of the representation theorem.

There is one rather annoying complication in giving a 'nominalistic definition' of (7), and that is that we may have inconveniently chosen a point b_2 which has either the highest or the lowest temperature of any point in the universe. So that we can save this complication for the end, let (7*) be the conjunction of (7) with the assertion that $\exists c \exists d$ (b_2 is strictly Scal-between c and d); we first give a 'nominalistic definition' of (7*), and then show how it can be used to obtain a 'nominalistic definition' of (7).

The idea of the nominalistic definition of (7*) is as follows. Take any two points c and d, on opposite sides of b_2 but as close to b_2 as one likes. Then (7*) says that there should be points y and z on opposite sides of x on a line L through x that is parallel to the line through a_1 and a_2, such that if you choose representation functions Φ and Ψ and a coordinate function Φ_L for L that is compatible with Φ, then for all points t on L other than x that are between y and z,

$$\frac{\Psi(t) - \Psi(x)}{\Phi_L(t) - \Phi_L(x)}(\Phi_L(a_2) - \Phi_L(a_1))$$

is exclusively between $\Psi(c) - \Psi(b_1)$ and $\Psi(d) - \Psi(b_1)$, i.e. is within a small amount of $\Psi(b_2) - \Psi(b_1)$. This is clearly what the usual platonistic explanation of (7*) amounts to.

Putting this more formally, and inserting both a clause to cover the case where $a_1 = a_2$ (which was implicitly excluded in giving the intuitive idea) and a clause asserting our temporary assumption about the non-extremality of b_2, we get

(8*) $\exists c \exists d$ (b_2 is strictly Scal-between c and d), and if $a_1 = a_2$ then $b_1 \approx_{Scal} b_2$, and if $a_1 \neq a_2$ then:

∀c∀d (if b_2 is strictly Scal-between c and d then there are points y and z such that yz Par a_1a_2 and x is strictly st-between y and z, and for all points t other than x that are strictly st-between y and z, (a_1a_2xt) E-Bet$_{st,Scal}$ $(xtb_1c)(xtb_1d)$.

(yz Par a_1a_2 means that the line segment $\overline{yz}$ is parallel to $\overline{a_1a_2}$ with $y \neq z$ and $a_1 \neq a_2$. It is defined in terms of st-Bet in note 32.) The previous paragraph should make it clear that (8*) is an adequate nominalistic definition of (7*): that is, it should be clear that it is platonistically provable that for any model of the joint axiom system and any representation functions Φ and Ψ, (8*) holds in the model if and only if (7*) is true.

What then are we to do if we want to nominalistically define not (7*) but (7)? One possibility is to recall that by cutting the size of a vector in half you cut the directional derivative with respect to that vector in half, so that if a_3 is on the line from a_1 to a_2 and halfway between them, and if b_3 is midway in temperature between b_1 and b_2, then the directional derivative with respect to $\Phi(a_2) - \Phi(a_1)$ will equal $\Psi(b_2) - \Psi(b_1)$ if and only if the directional derivative with respect to $\Phi(a_3) - \Phi(a_1)$ equals $\Psi(b_3) - \Psi(b_1)$. And what follows the 'if and only if' is explicated nominalistically by (8*), for the point b_3 is guaranteed to be strictly Scal-between two other points, namely b_1 and b_2. At least, this is guaranteed when it is not the case that $b_1 \approx_{Scal} b_2$; taking account of the possibility that $b_1 \approx_{Scal} b_2$ as well, the above remarks give us the nominalistic definition of (7):

(8) Either $b_1 \approx_{Scal} b_2$ and $\exists b\{$(8*) with b_1 and b_2 replaced by $b\}$, or not($b \approx_{Scal} b_2$) and $\exists a_3 \exists b_3 [a_3$ st-Bet $a_1 a_2$ and $a_1 a_3$ Par-Cong $a_3 a_2$ and $b_1 b_3$ Scal-Cong $b_3 b_2$ and $\{$(8*) with b_2 replaced by $b_3\}]$.

(Par-Cong is defined in note 32.) This is an adequate nominalistic definition of (7). Let us abbreviate it as D(x, a_1, a_2, b_1, b_2).

The reader may object that a nominalistic definition of (7) isn't enough: for since the range of our scalar may not exhaust the real numbers, it may happen that the directional derivative of T with respect to $\Phi(a_2) - \Phi(a_1)$ exists at $\Phi(x)$ but is too big to be represented in the form $\Psi(b_2) - \Psi(b_1)$. Actually, however, this causes no problem. For the directional derivative with respect to a vector exists at a point if and only if the directional derivative with respect to a shortened vector pointing in the same direction exists at that point; and by shortening sufficiently much, we can always get the value of the directional derivative to be as small as some actual temperature-difference. So, to say that the directional derivative with respect to $\Phi(a_2) - \Phi(a_1)$ exists at $\Phi(x)$, we need merely say that $\exists a_2' \exists c \exists d [a_2'$ st-Bet $a_1 a_2$ and (if $a_2 \neq a_1$ then $a_2' \neq a_1$) and $D(x, a_1, a_2', c, d)]$. This doesn't enable us to equate the value of the directional derivative with respect to $\Phi(a_2) - \Phi(a_1)$ to any actual temperature difference, but we never need to do that: we can always make do with stating the value of the directional derivative with respect to some smaller parallel vector, $\Phi(a_2') - \Phi(a_1)$, since cutting the size of the vector by some fraction always cuts the size of the directional derivative by the same fraction.

We can not only express the existence of the partial derivatives of T at a point $\Phi(x)$, but we can also express the differentiability of T at $\Phi(x)$ (i.e. the existence of a linear transformation that approximates T at $\Phi(x)$): the intuitive idea is that T is differentiable at $\Phi(x)$ if and only if

(a) for each a_1 and a_2, the directional derivative of T with respect to $\Phi(a_2) - \Phi(a_1)$ exists at $\Phi(x)$;

(b) for each a_1, a_2, a_3, a_4 that form a parallelogram with a_3 opposite a_1, and such that a_2, a_3, and a_4 are all close enough to a_1 so that the values of the directional derivatives of T with respect to $\Phi(a_2) - \Phi(a_1)$, $\Phi(a_3) - \Phi(a_1)$, and $\Phi(a_4) - \Phi(a_1)$ are all within the range of the temperature scale: the directional derivative with respect to $\Phi(a_3) - \Phi(a_1)$ is the sum of the directional derivatives with respect to $\Phi(a_2) - \Phi(a_1)$ and $\Phi(a_4) - \Phi(a_1)$.

It should be clear how to express this nominalistically given what has already been said.

E. Second (and Higher) Derivatives

Second derivatives of scalar fields are also no difficulty: we merely need to express the result of taking a first derivative by means of betweenness and congruence predicates, and apply the whole process again. (And we can reiterate still further to get third and higher derivatives.) More specifically, suppose we define D-Bet$(a_1, a_2; x, y, z)$ as:

$\exists a_2', b, c, d, e[a_2'$ st-Bet $a_1 a_2 \wedge (a_2 \neq a_1 \rightarrow a_2' \neq a_1) \wedge D(x, a_1, a_2', b, c)$
$\wedge D(y, a_1, a_2', b, d) \wedge D(z, a_1, a_2', b, e) \wedge c$ Scal-Bet $de]$;

this says that the directional derivative of T with respect to $\Phi(a_2) - \Phi(a_1)$ at $\Phi(x)$ is between the directional derivatives with respect to the same vector at $\Phi(y)$ and $\Phi(z)$. (a_2' is invoked to avoid the difficulties raised in the next to last paragraph of the preceding section.) Suppose we also define D-Cong($a_1, a_2; x, y, z, w$) as:

$\exists a_2', b, c, d, e, f [a_2'$ st-Bet $a_1 a_2 \wedge (a_2 \neq a_1 \rightarrow a_2' \neq a_1) \wedge D(x, a_1, a_2', b, c) \wedge D(y, a_1, a_2', d, c) \wedge \{(D(w, a_1, a_2', b, e) \wedge D(z, a_1, a_2', d, e)) \vee (D(w, a_1, a_2', e, b) \wedge D(z, a_1, a_2', e, d))\}]$;

this says (see Figure 6) that if we take the directional derivative of T with respect to $\Phi(a_2) - \Phi(a_1)$ at x, y, z, and w, the absolute value of the difference between this derivative at x and at y equals the absolute value of the difference between the derivatives at z and at w.

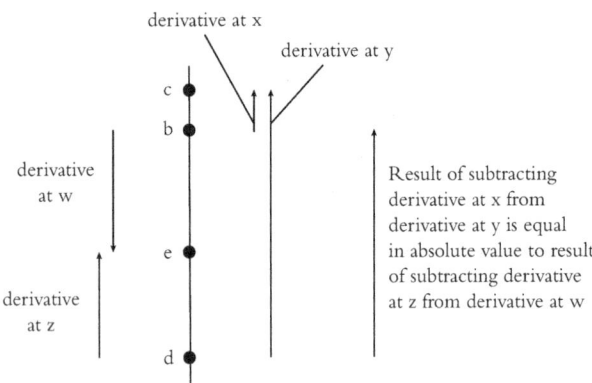

Figure 6

Finally, if the original system was an *ordered* scalar field, we can define D-Less ($a_1, a_2; x, y$) as:

$\exists a_2', b, c, d [a_2'$ st-Bet $a_1 a_2 \wedge (a_2 \neq a_1 \rightarrow a_2' \neq a_1) \wedge D(x, a_1, a_2', b, c) \wedge D(y, a_1, a_2', b, d) \wedge c$ Scal-Less $d]$.

Then for any fixed a_1 and a_2 the predicates D-Cong ($a_1, a_2; x, y, z, w$) and D-Bet ($a_1, a_2; x, y, z$) satisfy the axiom system for an unordered scalar field;

and if the original system was an ordered scalar field, then D-Cong, D-Bet, and D-Less obey the axiom system for an ordered scalar field. Consequently, we can think of these predicates as representing (for fixed a_1 and a_2) a scalar function, and differentiate with respect to a new vector $\Phi(a_4) - \Phi(a_3)$ as before. In doing so we get a new formula $E(x, a_1, a_2, a_3, a_4, y, z)$, with the representation theorem that (for any representation functions Φ and Ψ) this holds if and only if the second directional derivative of $\Psi \circ \Phi^{-1}$ with respect to $\Phi(a_2) - \Phi(a_1)$ and $\Phi(a_4) - \Phi(a_3)$ in that order exists at $\Phi(x)$ and has a value there equal to that of the first directional derivative of $\Psi \circ \Phi^{-1}$ with respect to $\Phi(a_2) - \Phi(a_1)$ at z minus the same first directional derivative at y.

Then the claim

$$\exists y, z, c[E(x, a_1, a_2, a_3, a_4, y, z) \wedge D(y, a_1, a_2, c, b_1) \wedge D(z, a_1, a_2, c, b_2)]$$

is sufficient for the second directional derivative with respect to $\Phi(a_2) - \Phi(a_1)$ and $\Phi(a_4) - \Phi(a_3)$ in that order to exist at $\Phi(x)$ and have a value there equal to $\Psi(b_2) - \Psi(b_1)$. Though sufficient, it isn't quite necessary, because of the fact that the range of the first derivative might not be big enough; but again it is a boring exercise to use linearity of derivatives to provide an emendation that is necessary and sufficient. Call this emended version $D^{(2)}(x, a_1, a_2, a_3, a_4, b_1, b_2)$. This is the desired formula for second directional derivatives; we can also express the existence of the second derivative as a bilinear operator (which will entail that we don't have to worry about the order of the vectors $\Phi(a_2) - \Phi(a_1)$ and $\Phi(a_4) - \Phi(a_3)$ in taking second directional derivatives) by the same method used to get the first derivative as a linear operator.[44]

---

[44] This section should suggest to those familiar with tensor methods something about how the nominalistic treatment of covector and cotensor fields and their differentiation is going to work (in the flat affine space-time we've been discussing). In flat space-time a contra-vector is represented as simply a pair of points, and covector and cotensor fields are treated by predicates that have slots to be filled by contra-vectors. In a general treatment one will represent a cotensor field of rank n by a betweenness predicate, a congruence predicate, and perhaps also a 'less than or equal to' predicate (depending on whether one needs order or not): these will have $3(2n+1)$, $4(2n+1)$, and $2(2n+1)$ places respectively, where each of the $(2n+1)$-place units represent the endpoints of n contravectors plus a point at which the field is being evaluated. (In dealing with differentiation it wasn't necessary to use all of these places, because of the fact that the cotensor resulted from an independently given scalar field.)

In developing gravitational theory nominalistically it is possible to take as one's primitive a predicate representing the gravitational field intensity covector, rather than the gravitational

F. Laplaceans

If we now particularize the discussion from arbitrary affine 4-dimensional space-times to the Newtonian space-time considered earlier, we can make statements about the Laplacean of a given scalar field. For example, we can say that the Laplacean at a point x exists and is zero: to say this is to say that the field is twice differentiable at x and (see Figure 7) that there is an st-basic region R containing x such that[45] for all a, b, c in R that are simultaneous to x, if $\overline{xa}$, $\overline{xb}$, and $\overline{xc}$ all have the same length and are pairwise orthogonal, then there are points d, e, and f such that:

> the second directional derivative with respect to $\Phi(a) - \Phi(x)$ taken twice (i.e. with respect to $\Phi(a) - \Phi(x)$ and $\Phi(a) - \Phi(x)$) is equal to $\Psi(e) - \Psi(d)$;
>
> the second directional derivative with respect to $\Phi(b) - \Phi(x)$ taken twice is equal to $\Psi(f) - \Psi(e)$; and
>
> the second directional derivative with respect to $\Phi(c) - \Phi(x)$ taken twice is equal to $\Psi(d) - \Psi(f)$.

(Orthogonality for purely spatial vectors, i.e. vectors whose endpoints are simultaneous, is definable in terms of the spatial congruence relation.)

This explains the claim that the Laplacean is zero in terms of other claims that we have already seen how to express nominalistically (and without appeal to any non-invariant notions).

---

potential scalar. This is in fact a more natural approach in some respects, but though it isn't ultimately any more complicated than the approach given here, it seemed to me that the approach given here would be conceptually less demanding as an introduction to the kind of nominalistic methods I'm using.

I believe that the ideas here are extendible to curved space-time. One natural approach to doing this would be to take contravectors at a point as geodesic segments emanating from that point and contained in a convex normal neighborhood of that point. (See pp. 33–4 of Hawking and Ellis 1973 for a definition of a convex normal neighborhood, and a sketch of a proof that every point of a manifold lies in such neighborhoods and that within any such neighborhood there is a natural diffeomorphism between geodesic segments on the one hand and contravectors characterized in terms of a tangent space on the other.) Cotensors would then be treated by predicates of contravectors, in the manner of two paragraphs back; contratensors of rank greater than 1 need not be treated directly, since the effect of them can be gotten by 'index raising' which can be done by the method of section I of this chapter. It ought to be possible, by a parallel transport predicate, to describe space-time curvature and to develop differential geometry. It is not however a trivial task to work out the details of this, for the whole construction would have to be based on a representation theorem of a more complicated kind than any I have seen.

[45] R is invoked to keep the second directional derivatives small enough.

LAPLACEANS 77

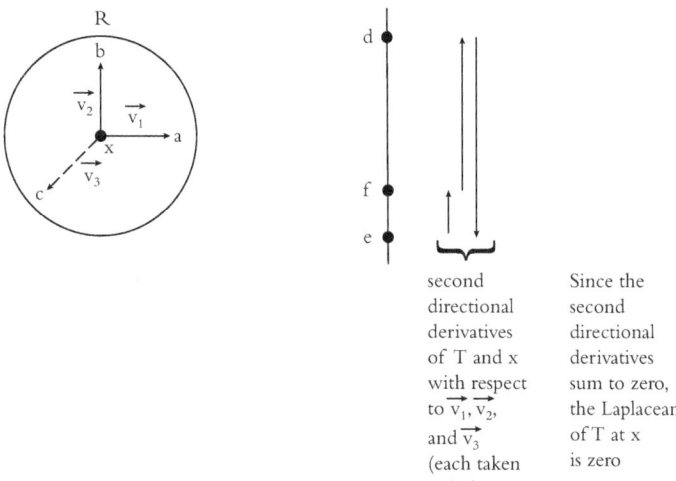

second directional derivatives of T and x with respect to $\vec{v_1}$, $\vec{v_2}$, and $\vec{v_3}$ (each taken twice)

Since the second directional derivatives sum to zero, the Laplacean of T at x is zero

Figure 7

We can also say that the ratio[46] of the Laplacean at x to the Laplacean at x' is equal to the ratio of the difference in scalar values between p and q to the difference in scalar values between p' and q': we simply say (cf. Figure 8) that there are a, b, c simultaneous to x, and a', b', and c' simultaneous to x', such that $\overline{xa}$, $\overline{xb}$, $\overline{xc}$ $\overline{x'a'}$, $\overline{x'b'}$, and $\overline{x'c'}$ all have the same length, and that the first three are pairwise orthogonal and the last three are too, and such that for some t, u, t', u':

(a) the second directional derivative at x with respect to $\Phi(a) - \Phi(x)$ taken twice is $\Psi(t) - \Psi(q)$; with respect to $\Phi(b) - \Phi(x)$ taken twice is $\Psi(u) - \Psi(t)$; and with respect to $\Phi(c) - \Phi(x)$ taken twice is $\Psi(p) - \Psi(u)$;
(b) the same as (a) but using primed points (a', x', t', p', etc.).

---

[46] I interpret this and other ratio statements as merely a convenient abbreviation of the corresponding product statement: i.e.

$$\frac{\alpha}{\beta} = \frac{\gamma}{\delta}$$

simply abbreviates

$$\alpha\delta = \beta\gamma.$$

This convention about the meaning of ratio statements enables us to avoid boring qualifications about the cases where $\beta = 0$ or $\delta = 0$.

78 GRAVITATIONAL THEORY NOMINALIZED

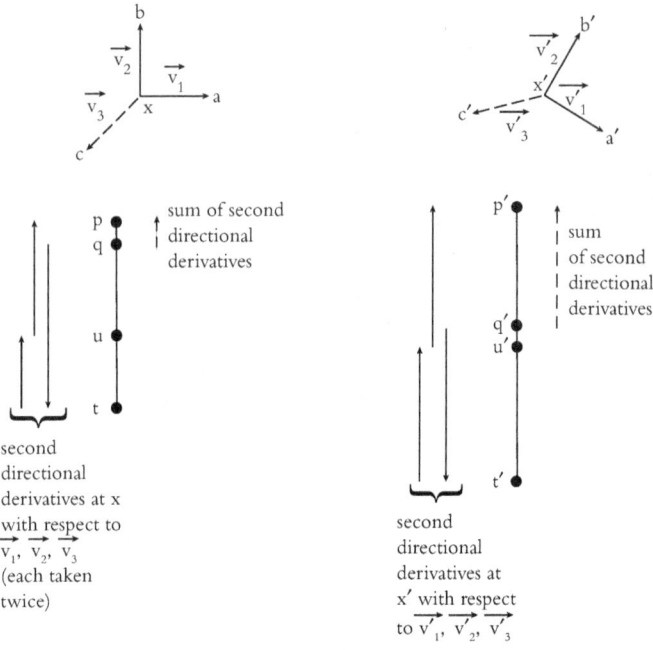

Ratio of dotted arrows represents ratio of Laplaceans

Figure 8

(Again I have stated this in a platonistic way, using Φ, Ψ, etc.; but I have used these platonistic devices only in contexts that we have already seen how to nominalize.)

Finally, if we are dealing with an ordered scalar field, then we can say that the Laplacean has a value less than or equal to zero at x, by an obvious modification of how we said that it had value zero at x.

We can also make slightly more complicated invariant statements about the Laplacean nominalistically and without appeal to non-invariant entities, but the three statements described above will suffice for Newtonian gravitational theory.

G. Poisson's Equation

By the Newtonian theory of gravitation I mean the theory of motion for an arbitrary particle, assuming that the only forces acting on the particle are gravitational forces. Given the space-time framework, which I have already shown how to handle nominalistically, the Newtonian theory of gravitation can be stated in two laws: a field equation governing a certain scalar field (the gravitational potential), and an equation of motion. The field equation is Poisson's equation, which says that at any point the Laplacean of the gravitational potential is proportional to the mass-density at that point, the proportionality constant being negative. (The absolute value of the proportionality constant has no invariant significance within this theory: to give it significance you have to impose independent constraints on the mass scale and on the scales of other quantities.)

For the moment let's forget about the requirement that the proportionality constant be negative, and require only that it be non-zero. Then the field equation can be restated as the conjunction of two claims:

(9a) at any point the Laplacean of the gravitational potential is zero if and only if the mass density at that point is zero;

(9b) at any two points at which the mass density is not zero, the ratio of the Laplaceans of the gravitational potential is equal to the ratio of the mass-densities.

Obviously the preceding discussion gives us most of the machinery required for saying this. All that is missing is that I haven't yet talked about the proper way to treat mass-density. Mass-density is a scalar field of a rather special sort: a symptom of the special nature of this field is that its scale is 'less arbitrary' than the scale for gravitational potential, i.e. it is a ratio scale (or a log-interval scale)[47] rather than an interval scale. The special nature of the scale means that a proper axiomatization of mass-density would involve a more complicated set of primitives and axioms than the ones suggested above for scalar fields generally. Nonetheless, the primitive used above for ordered scalar fields would be included among or definable from the primitives used in the more complicated treatment, and the axioms alluded to above for ordered scalar fields would be included among or derivable from the axioms in the more complicated treatment. Another thing definable from the primitives is the notion of having a mass-density of zero; and if we assume that there

---

[47] For a discussion of log-interval scales and of why density should be regarded as a log-interval scale rather than a ratio-scale, see Krantz et al. 1971: 10–11, 484–7.

are points of mass-density zero, we can avoid having to consider the details of the proper treatment of the mass-density scale in nominalizing Poisson's equation. For then we can express the density ratio between points x and y as a ratio between the density differences $\rho(x) - \rho(z)$ and $\rho(y) - \rho(z)$, where z is a point at which the mass-density is zero; and we can compare ratios of density differences with ratios of differences in gravitational potentials, by the device used earlier to compare the latter ratio with ratios of oriented spatio-temporal intervals on a line in section C of this chapter. Since our invariant treatment of the Laplacean gives us a way to compare ratios of Laplaceans of the gravitational potential with ratios of differences of gravitational potential, we can put these things together to compare ratios of Laplaceans to ratios of densities. That is, roughly (i.e. ignoring complications arising from the possible finiteness of the gravitational potential scale, which we know by now how to handle), (9b) is equivalent to:

for any points x and x' and any point z at which the mass density is zero and any points p, q, p', and q':

$$\frac{\rho(x) - \rho(z)}{\rho(x') - \rho(z)} = \frac{\Psi(q) - \Psi(p)}{\Psi(q') - \Psi(p')}$$

if and only if

$$\frac{\text{the Laplacean of } \Psi \circ \Phi^{-1} \text{ at } x}{\text{the Laplacean of } \Psi \circ \Phi^{-1} \text{ at } x'} = \frac{\Psi(q) - \Psi(p)}{\Psi(q') - \Psi(p')};$$

and all the notions appearing in this are ones we have previously seen how to define nominalistically. The appeal to points at which the density is zero in this treatment is a bit inelegant, but it can be avoided on a fuller treatment that takes more seriously the fact that density has a ratio (or more accurately, log-interval) scale.[48]

---

[48] On the fuller treatment we can state and prove that for any points a and b there are points c, d, e, and f such that the ratio of the mass at a to the mass at b equals the ratio of the difference in mass between c and d to the difference in mass between e and f. (Using the ratio convention of note 46, and also the (obviously true) assumption that the mass density is not the same at each point. If for some reason one wants to avoid that assumption, the case of uniform mass density can be treated as a separate case.) The assumption in the text that there are points at which the mass density is zero is really just a simple way of getting this result.

The only thing that remains to be done in treating Poisson's equation is to express the fact that the proportionality constant in the equation is negative. Here for the first time we must use the primitive 'Scal-Less', for Poisson's equation is *not* invariant under reversal of signs of the gravitational potential: a world in which Poisson's equation held with a positive proportionality constant would be a world where objects had more gravitational potential energy at the surface of the earth than on a mountain top. It is clear, however, that the fact that the constant is negative amounts merely to the fact that

(9c) the Laplacean of the gravitational potential is always less than or equal to zero,

and we've already seen how to express this. So the nominalization of Poisson's equation is complete.

H. Inner Products

It now remains only to give nominalistically the law of motion for Newtonian gravitational theory: this says that the acceleration of a point-particle subject only to gravitational forces is at each point on the particle's trajectory equal to the gradient of the gravitational potential at that point.[49] The invariant content of this law is exhausted by the claim that the gradient is proportional to the acceleration, with a positive proportionality constant that is the same for all trajectories. (It is usual to use a proportionality constant of 1, but this simply reflects an arbitrary choice of scale for the gravitational potential, relative to scales for spatial distance and for time.)

To state the law nominalistically, we first need to be able to compare ratios of inner products of purely spatial vectors with ratios of scalar differences (where a purely spatial vector is one whose endpoints are simultaneous). But doing this is easy, given what we've done so far: first, given four simultaneous points x_1, x_2, y_1, and y_2 (cf. Figure 9(a)) let z_1 be chosen so that $\overrightarrow{x_1 z_1}$ is parallel to $\overrightarrow{y_1 y_2}$, and points in the same direction as it, and has the same length as it; and let z_2 be the point at which the perpendicular to the line $x_1 x_2$ through z_1 meets $x_1 x_2$. (If $x_1 = x_2$, so that 'the

---

[49] Point-particles are presumably an idealization, and an idealization that gives rise to some difficulties; but these difficulties arise on the usual platonistic field-theoretic formulations of physics too, and hence don't seem specially relevant to the issue of nominalism.

82 GRAVITATIONAL THEORY NOMINALIZED

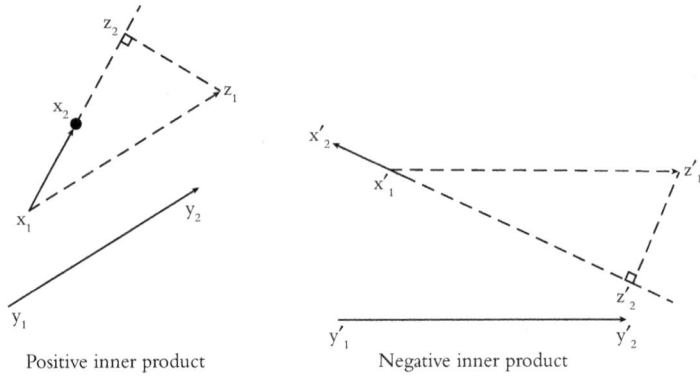

Positive inner product   Negative inner product

Figure 9(a)

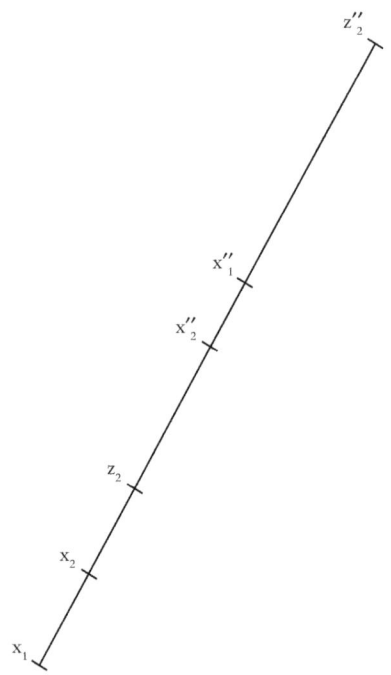

Figure 9(b)

line x_1x_2' isn't unique, it won't matter where you take z_2.) All this is nominalistically describable, of course. If z_2 is x_1 or x_2 is x_1, the inner product of $\overrightarrow{x_1x_2}$ and $\overrightarrow{y_1y_2}$ is zero; if x_1 is strictly between x_2 and z_2, the inner product is negative; otherwise, the inner product is positive. Similarly, given simultaneous points x'_1, x'_2, y'_1, and y'_2, we can construct z'_2 in a similar way, and again its location determines the sign of the inner product.

So much for the signs of the inner products $\overrightarrow{x_1x_2} \cdot \overrightarrow{y_1y_2}$ and $\overrightarrow{x'_1x'_2} \cdot \overrightarrow{y'_1y'_2}$; now how about about their magnitudes? Intuitively, the ratio of the magnitudes of the absolute values of these inner products is equal to the length of $\overrightarrow{x_1x_2}$ times the length of $\overrightarrow{x_1z_2}$ divided by the length of $\overrightarrow{x'_1x'_2}$ times the length of $\overrightarrow{x'_1z'_2}$. But this intuitive idea has to be put nominalistically. To do so, we first make a copy of the triple $\langle x'_1, x'_2, z'_2 \rangle$ on the line through x_1, x_2, and z_2 (see Figure 9(b)): let x''_1, x''_2, and z''_2 be on the same line as x_1, x_2, and z_2, with $x''_1x''_2$ S-Cong $x'_1x'_2$ and $x''_1z''_2$ S-Cong $x'_1z'_2$ and $x''_2z''_2$ S-Cong $x'_2z'_2$. (Spatial congruence makes sense in this context, since x'_1, x'_2, and z'_2 are all simultaneous and so are x''_1, x''_2, and z''_2.) Now suppose u_1, u_2, u'_1, and u'_2 are four further points (not necessarily simultaneous) at which we are interested in a scalar property like gravitational potential. Then to say that

$$\frac{\overrightarrow{x_1x_2} \cdot \overrightarrow{y_1y_2}}{\overrightarrow{x'_1x'_2} \cdot \overrightarrow{y'_1y'_2}} = \frac{\Psi(u_1) - \Psi(u_2)}{\Psi(u'_1) - \Psi(u'_2)}$$

(where the dots indicate inner product, and the convention on ratio-statements stated in note 46 is in force), is simply to say that

$$(\Phi_L(x_2)-\Phi_L(x_1))(\Phi_L(z_2)-\Phi_L(x_1))(\Psi(u'_2)-\Psi(u'_1)) = \\ (\Phi_L(x''_2)-\Phi_L(x''_1))(\Phi_L(z''_2) - \Phi_L((x''_1))(\Psi(u_2) - \Psi(u_1))$$

(where Φ_L is a coordinatization of the line L that's compatible with Φ, and $x_1, x_2, z_2, x''_1, x''_2$, and z''_2 all reside on L); and we saw how to say this nominalistically at the end of section C.

To be more precise, the above sketch shows how to state a formula governed by the representation theorem that for any Φ and Ψ, the formula holds if and only if

(10) $\overrightarrow{x_1x_2}, \overrightarrow{y_1y_2}, \overrightarrow{x'_1x'_2}$ and $\overrightarrow{y'_1y'_2}$ are purely spatial vectors, and

$$\frac{\overrightarrow{x_1x_2} \cdot \overrightarrow{y_1y_2}}{\overrightarrow{x'_1x'_2} \cdot \overrightarrow{y'_1y'_2}} = \frac{\Psi(u_1) - \Psi(u_2)}{\Psi(u'_1) - \Psi(u'_2)}$$

I. Gradients

The availability of the notion of inner product ratios in Newtonian space-time allows us to associate with each 'ratio of derivative operators' a 'ratio of vectors': namely, the 'ratio' of the gradient vectors that correspond to the derivative operators. (This then is the analog of 'index raising' in tensor analysis.) To be more precise, observe that since we can express (10), we can also express

(11) $\overrightarrow{x_1 x_2}, \overrightarrow{y_1 y_2}, \overrightarrow{x'_1 x'_2}$ and $\overrightarrow{y'_1 y'_2}$ are purely spatial, and

$$\frac{\overrightarrow{x_1 x_2} \cdot \overrightarrow{y_1 y_2}}{\overrightarrow{x'_1 x'_2} \cdot \overrightarrow{y'_1 y'_2}}$$

is equal to the ratio of the directional derivative of $\Psi \circ \Phi^{-1}$ with respect to $\overrightarrow{y_1 y_2}$ at z to the directional derivative of $\Psi \circ \Phi^{-1}$ with respect to $\overrightarrow{y'_1 y'_2}$ at z';

for we know how to nominalistically compare directional derivatives with scalar-difference ratios like

$$\frac{\Psi(u_1) - \Psi(u_2)}{\Psi(u'_1) - \Psi(u'_2)}$$

and so we can certainly compare *ratios* of directional derivatives with such scalar-difference ratios.[50] But then we can say that (11) holds for *all* purely spatial vectors $\overrightarrow{y_1 y_2}$ and $\overrightarrow{y'_1 y'_2}$; and this is in effect to say

(12) there is a real number k such that $\overrightarrow{x_1 x_2}$ is k times the gradient of $\Psi \circ \Phi^{-1}$ at z and $\overrightarrow{x'_1 x'_2}$ is k times the gradient of $\Psi \circ \Phi^{-1}$ at z'.

We would also like to be able to say that (12) holds with some positive proportionality constant k. Call this claim (12'). To say this, we simply add to the nominalistic definition of (12) the further claim that for all purely spatial vectors $\overrightarrow{y_1 y_2}$, the inner product $\overrightarrow{x_1 x_2}$ and $\overrightarrow{y_1 y_2}$ at z is positive if and only if the directional derivative with respect to $\overrightarrow{y_1 y_2}$ at z is positive, and analogously for $\overrightarrow{x'_1 x'_2}$ and z'. (This involves a second use of the relation 'less than in gravitational potential'.)

---

[50] This requires a bit of care because of the fact that the range of the scalar may be a finite interval, but as usual the difficulty is resolved by cutting the size of the vectors with respect to which the directional derivatives are taken.

J. Differentiation of Vector Fields

(12′) is the key to a nominalistic understanding of the right-hand side of the law of motion; we now have to deal with the left-hand side, which involves the notion of the acceleration of a point particle. Until now, the ontology of the theory has included only space-time regions; now we are invoking new entities, viz. point particles, and to go with them we will need a new primitive predicate, that of a point-particle *occupying* a space-time point. From this we can define the notion of the *trajectory* of a particle: it is the region consisting of all space-time points that the particle ever occupies.[51]

An implicit presupposition of the law of motion is that the trajectory of each point particle is a region that is connected in the st-topology given in section A of this chapter and which contains no two simultaneous points. I will call any such region of space-time (whether there is a particle there or not) *trajectory-like*. The most important trajectory-like regions to consider, aside from actual trajectories, are those straight lines in space-time that are not purely spatial: as Newton's bucket argument more or less shows, such lines must play a crucial role in formulating the law of motion.

In formulating the law of motion, it is necessary to speak of the spatial separation between trajectory-like regions, and first and second derivatives of this spatial separation. Since the spatial separation between trajectory-like regions is in effect a vector—it has direction as well as magnitude—we need to extend the treatment of differentiation in section D to vector fields.[52]

More particularly, let S and T be any two trajectory-like regions. It is useful heuristically to think of them as 'defining a vector field' as follows: for any point x, regard the vector field as being defined at x if and only if both S and T contain points simultaneous to x; and regard the value of the field there as the vector whose initial point and terminal points are

---

[51] Actually, the *only* use of either particles or the notion of occupation in the theory is in defining the notion of trajectory. Consequently, one could if one liked avoid explicitly introducing particles, and take 'trajectory' instead of 'occupies' as primitive. I don't claim any philosophical significance for this; I note it only because it allows a slight technical simplification of the discussion in the next chapter.

[52] Contravariant vector fields, that is. Covariant vector fields have in effect been dealt with earlier: cf. note 44.

the points of S and T respectively that are simultaneous to x. Let a_1 and a_2 be any points. (In the cases of interest for this particular vector field, they will not be simultaneous.) What can be said, invariantly, about the value of the directional derivative of this vector field at x with respect to the vector $\overrightarrow{a_1 a_2}$? Well, directional derivatives of scalar fields could be objectively equated with differences of scalars, so you would expect that directional derivatives of vector fields could be objectively equated with differences of vectors. But a difference of vectors is itself a vector. What we should expect, then, is that we could define nominalistically a formula D-Vec(x, a_1, a_2, b_1, b_2), meaning intuitively that at x the directional derivative of the spatial separation of T from S with respect to $\overrightarrow{a_1 a_2}$ exists and is equal to the spatial separation of b_2 from b_1: we ought to be able to define this without assigning a number to the length of the spatial separation, or to anything else.

Doing this involves only a fairly straightforward generalization of what was done in section D; in fact, in some ways the definition of vector differentiation needed here is easier than the definition of scalar differentiation given there, since space-time unlike the range of temperature or gravitational potential is being assumed to be infinite in extent, so there's no need for all the fancy footwork involving the linearity of derivatives which the possibility of finiteness forced on us. But there are also some additional complications in the vector case. We proceed as follows.

In analogy with section C we can easily define a predicate $(xyzw) =_{st,st} (rstu)$ expressing the equality of signed products of distances on two distinct lines, i.e. the equality of $(\Phi_L(y) - \Phi_L(x)) \cdot (\Phi_{L'}(w) - \Phi_{L'}(z))$ and $(\Phi_L(s) - \Phi_L(r)) \cdot (\Phi_{L'}(u) - \Phi_{L'}(t))$ where x, y, r and s all lie on a single line L and the other points all lie on a single line L'. It's easy to generalize this to a predicate $(xyzw) =_{st,st}^{par} (rstu)$ where xy is merely required to be parallel to rs, and zw to be parallel to tu. A simple way to do this is to let Parallelogram*(x, y, z, w) mean that x, y, z, and w either are the vertices of a parallelogram with x opposite z, or they are the vertices of a limiting case of a parallelogram, that is, either $x = y$ and $z = w$ or $x = w$ and $z = y$. Then $(xyzw) =_{st,st}^{par} (rstu)$ can be defined (see Figure 10) as: there are points s' and u' such that Parallelogram*(x, s', s, r) and Parallelogram*(z, u', u, t) and $(xyzw) =_{st,st} (xs'zu')$.

We are now ready to define D-Vec (x, a_1, a_2, b_1, b_2). Let p_x and q_x be the points on S and T respectively that are simultaneous with x. (As remarked

DIFFERENTIATION OF VECTOR FIELDS 87

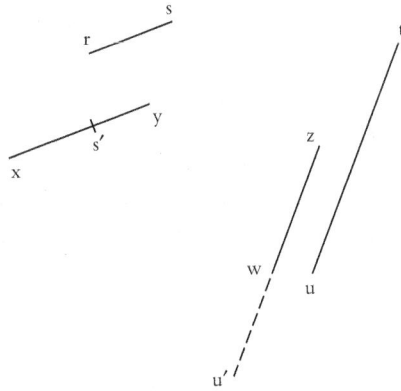

Figure 10

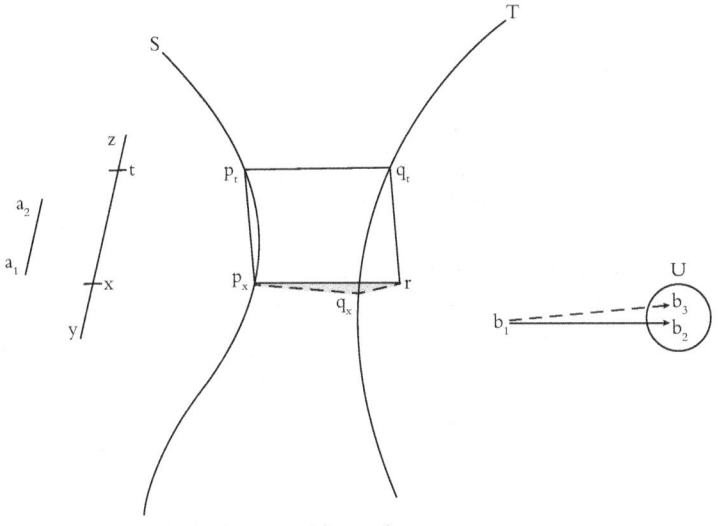

q_x is simultaneous with p_x and r,
but need not be in the p_x-p_r-q_t-r plane

Figure 11

above, there must be such for the vector field we're interested in to be defined at x.) Then D-Vec (x, a_1, a_2, b_1, b_2) comes to this (see Figure 11):

If $a_1 = a_2$ then $b_1 = b_2$; and if $a_1 \neq a_2$ then $\forall U$[if U is an st-basic region containing b_2, then there are y and z such that:

(a) yz Par $a_1 a_2$;

(b) x is strictly st-between y and z; and
(c) for any t other than x that is strictly st-between y and z, there is a point r such that Parallelogram* (p_x, p_t, q_t, r) and a point b_3 in U such that $(xtb_1b_3) = {}^{par}_{st,st}(a_1a_2q_xr)$.

The last clause says in effect that if $\overrightarrow{xt}$ is h times $\overrightarrow{a_1a_2}$ for some real number h, then $\overrightarrow{q_xr}$ (which is the change in spatial separation in passing from x to t) is h times $\overrightarrow{b_1b_2}$, and hence is 'close to' the vector $h \cdot \overrightarrow{b_1b_2}$. It is easy to see that if D-Vec(x, a_1, a_2, b_1, b_2) then b_1 and b_2 are simultaneous, which is what you'd expect since the endpoints of the spatial separation vector are always simultaneous; also, it is easy to see that when in addition a_1 is simultaneous to a_2 then $b_1 = b_2$, which again is what you'd expect since the spatial separation vector has the same value at all simultaneous points.

Observe that the derivative of a vector field with respect to a fixed vector is again a vector field; consequently we can immediately differentiate again with respect to the same or a different vector. Second and higher derivatives are thus a bit easier for vectors than for scalars, where it took a bit of work to put the result of taking the first derivative into the same format as the scalar field that we had started with.

K. The Law of Motion

The law of motion can now be stated in any of a variety of ways. Perhaps the most natural is to introduce the concept of a tangent to a trajectory (or a trajectory-like region) at a point: a tangent to a trajectory T at a point z is a straight line S through z, such that the directional derivative of the spatial separation between T and S with respect to any vector exists and is zero at z. The tangent to T at z is unique if it exists. Let us call a trajectory (or trajectory-like region) *differentiable at z* if it has a tangent at z and this tangent is not purely spatial. We can now take the preliminary part of the law of motion to consist in the claim that the trajectory of any point particle is both trajectory-like (in the sense defined in section J) and differentiable.

The main part of the law of motion requires that we compare the accelerations of points on the same or different trajectories with the gradients of the gravitational potential at those points. Let T and T' be any trajectories, and let z and z' be any points on them. Let S and S′ be the tangents to T and T′ at z and z' respectively, and let y and y' be points on S and S′ such that $\overrightarrow{zy}$ and $\overrightarrow{z'y'}$ are temporally congruent and have the same temporal orientation. The law of motion is then simply that (for any such T, T′, z, z', S, S′, y, and y') there is a positive real number k such that:

(a) the second directional derivative of the spatial separation of S from T at z with respect to $\overrightarrow{zy}$ taken twice is k times the gradient of the gravitational potential at z; and

(b) the second directional derivative of the spatial separation of S′ from T′ at z' with respect to $\overrightarrow{z'y'}$ taken twice is k times the gradient of the gravitational potential at z'.

But the previous two sections show how to say this: it is simply a matter of plugging the appropriate second directional derivatives into (12′). So the nominalistic formulation of the law of motion is complete, and this together with our previous nominalization of Poisson's equation gives us a complete nominalistic formulation of Newtonian gravitational theory.

L. General Remarks

Let us review the strategy we have followed. We started out by giving a joint axiom system containing axioms for space-time, axioms for the gravitational potential, and (though we didn't mention them until later) axioms for mass-density. We then proved that for any model of this joint axiom system, there would be a 1–1 function Φ from the domain of the model (i.e. the space-time points) onto $\mathbb{R}^4$, and two functions Ψ and ρ from the domain into the reals, all satisfying certain homomorphism conditions. We then showed that there were further nominalistic statements expressible using the primitives of the joint axiom system (together with the notion of a particle occupying a point) such that if these further nominalistic statements were true in the model then the usual platonistic formulation of Newton's theory of gravitation would come out true (taking Φ to be the spatio-temporal coordinate function, Ψ the gravitational potential function, and ρ the mass-density function).

So the nominalistic formulation of the physical theory in conjunction with standard mathematics yields the usual platonistic formulation of the theory; and conversely, the nominalistic formulation is a consequence of the platonistic formulation, given standard mathematics. From this it follows that any statement in the proposed nominalistic language that is a consequence of the platonistic axioms and standard mathematics is a consequence of the nominalistic axioms unaided by mathematics: for as we saw earlier in this monograph, mathematics when applied to nominalistic axiom systems does not yield any nominalistically statable conclusions you couldn't get otherwise. So the nominalistic formulation of physics and the platonistic formulation have precisely the same nominalistically statable consequences; and so mathematical entities are theoretically dispensable in the theory of gravitation.

I would like to conclude this chapter by saying that the nominalistic formulation of gravitational theory proposed here is not as complicated as it may look; or more accurately, that though it is complicated, that is because platonistic physics is complicated too, though it may be presented simply once we have explained a lot of platonistic apparatus (such as gradients, Laplaceans, and so forth). Most of what I have spent time doing in this chapter is to develop a nominalistic version of that complicated platonistic apparatus; and I doubt that my development of that is very much more complicated than the development of the analogous platonistic apparatus. Also, most of what I've developed is quite general: e.g. the treatment of differentiation in section D works for arbitrary functions from one affine space to another; and the treatment of gradients, Laplaceans, and inner products works for an arbitrary affine space on which there is a congruence relation defined, or which has distinguished subspaces (e.g. in the Newtonian example the 3-dimensional subspaces produced by factoring by the simultaneity relation) on which a congruence relation is defined. Once the development of this apparatus is complete, the laws of Newtonian gravitational theory can be presented very quickly in terms of it; again, probably almost as simply as they are presented on a platonistic approach. I believe that the reader who works through this material at all carefully will soon convince himself that this is so.

I do not of course claim that the nominalistic concepts are anywhere near as convenient to work in solving problems or performing computations: for these purposes, the usual numerical apparatus is a practical necessity. But it is a necessity that the nominalist has no need to forgo: he can treat the apparatus in the way suggested earlier in the book, i.e. as a useful instrument for making deductions from the nominalistic system that is ultimately of interest; an instrument which yields no conclusions not obtainable without it, but which yields them more easily.

9
Logic and Ontology

In this final chapter I want to deal with worries that the reader might have as to whether the version of physics presented here is genuinely nominalistic. Some of these worries, about my realist attitude toward space-time, seem to me entirely misguided; I have discussed this matter early in Chapter 4. The remaining worry, which I take much more seriously, stems from the fact that there are two respects in which I have overstepped the bounds of first-order logic. The fact that I have overstepped these bounds raises two questions:

(a) What are the prospects for making do with first-order logic?
(b) If the prospects are poor, what impact will this have on nominalism?

Although I strongly suspect that one can make do with first-order logic in developing gravitational theory nominalistically, proving this would involve much more work than proving the adequacy of the second-order nominalistic system considered in previous chapters. Consequently I will begin by considering question (b). Afterwards, I will say a little bit about question (a), for this question is of interest whatever the answer to (b).

The two respects in which I have overstepped the bounds of first-order logic are:

(i) that in axiomatizing the geometry of space-time and the scalar orderings of space-time points, I have invoked what I called in Chapter 4 'the complete logic of the part/whole relation' or 'the complete logic of Goodmanian sums'; and
(ii) that in comparing products of intervals in section B of Chapter 8, I have invoked the binary quantifier 'fewer than'.

Now strictly speaking, we do not really need (ii) in addition to (i): the logic of Goodmanian sums is sufficient to give us the cardinality comparisons we need as well as the representation theorems we need. (See note 66.) But I think that for a variety of reasons it is heuristically advantageous to keep the use of extra logic in making cardinality comparisons separate from the use of extra logic in giving representation theorems, so I will put no weight on the fact that one can make the cardinality comparisons one needs using only the logic of Goodmanian sums.

I

Let us introduce the symbol $\mathcal{F}$ for the binary quantifier 'fewer than': that is, let '[Apple (x)] $\mathcal{F}x$ [Orange (x)]' mean 'there are fewer apples than oranges'. Now, if we think of the full 'fewer than' quantifier as making discriminations among infinite cardinalities (e.g. as such that 'there are fewer points than regions' is true), then there is no need to invoke the full 'fewer than' quantifier in our theory: we can invoke a slightly weaker quantifier $\mathcal{F}_0$ which makes no distinction between infinite cardinalities. (Thus '[Apple (x)] $\mathcal{F}x$ [Orange (x)]' and '[Apple (x)] $\mathcal{F}_0 x$ [Orange (x)]' have the same truth-value if there are only finitely many oranges; and if there are infinitely many oranges then even if the number of oranges is uncountable, '[Apple (x)] $\mathcal{F}_0 x$ [Orange (x)]' is true if and only if there are only finitely many apples.) In the future, I will use only the quantifier $\mathcal{F}_0$, and will use the term 'fewer than' in accordance with the meaning of $\mathcal{F}_0$ rather than with $\mathcal{F}$ on those occasions where it matters.

It will turn out actually that in Newtonian gravitational theory, if we add a new predicate to the theory then the 'fewer than' quantifier can very easily be dispensed with in favor of a still simpler quantifier, the binary quantifier '$\exists_{fin}$' meaning 'there are only finitely many' (i.e. '$\exists_{fin} x$ Apple (x)' means 'there are only finitely many apples'). So it is really this quantifier that raises the issues of whether cardinality comparisons like those made in section B of Chapter 8 are nominalistically legitimate.

The first point I want to make is that use of the 'finitely many' quantifier does not seem pretheoretically to involve one in ontological cornmitments to abstract entities. If I assert or deny that there are only finitely many grains of sand, this appears to involve no commitment whatever to abstract entities, just as it appears to involve no commitment to abstract entities to assert or deny that there are less than 87 grains of sand. With regard to assertions or denials that there are less than 87 grains of sand, three attitudes are possible. First one could say that $\exists_{<87}$ (and/or its denial $\exists_{\geq 87}$) is simply a part of logic, as are $\exists_{\geq 1}$, $\exists_{\geq 2}$, etc.: on this view, logic includes not only first-order logic, but the recursive set of axioms for $\exists_{\geq 1}$, $\exists_{\geq 2}$, etc. given in Chapter 2. Second, one could say that despite appearances, 'there are at least 87 grains of sand' does involve ontological commitments to abstract entities: for, one might say, what it means is that there is a 1-1 function from the set $\{0, 1, \ldots, 86\}$ into the set of all grains of sand, and hence involves a commitment to functions, numbers, and sets as well as grains of sand. This view is quite implausible—e.g. it makes the sentence 'There are at least 87 grains of sand, but there are no numbers, functions or sets' inconsistent! Intuitively, the claim about the 1-1 function seems intimately related to the 'there are at least 87' claim—it is what I've called an *abstract counterpart* of the claim—but though intimately related, the claims are distinct. The third possible view is that the 'there are at least 87' claim is equivalent in meaning to a claim in first-order logic with identity. As a view about meaning this isn't really very plausible; but on the first view too 'there are at least 87 grains of sand' is *logically equivalent to* a claim in first-order logic with identity, so unless we are very hung up on meaning then the third view and the first do not seem importantly different in this case.

Now let's consider the claim 'there are only finitely many grains of sand', or its denial. Here one can not take the third of the lines used for 'there are fewer than 87 grains of sand' or its denial: the claim about finitude cannot be identified (even up to logical equivalence) with a claim in first-order logic plus identity. But the first line is still possible: we can take 'there are only finitely many' as a primitive quantifier, just as we could take 'there are at most 87' as primitive.[53] Admittedly, doing this has some consequences that are not entirely attractive: if we take this as a quantifier and also define logical consequences *à la* Tarski 1936, then the consequence relation is neither compact nor recursively enumerable. But the only alternative to taking 'there are only finitely many' as a primitive quantifier (one we might or might not elect to use in our theories) seems to be to say that

---

[53] A truth theory for a language containing '$\exists_{fin}$' would of course have to use the notion of finiteness. But that is no objection to the clarity of '$\exists_{fin}$' or the legitimacy of regarding it as logical, any more than the fact that the clause in a truth theory for the standard existential quantifier uses the notion of existence shows that that quantifier isn't clear or isn't part of logic.

despite all appearances 'there are infinitely many grains of sand' commits you to the existence of functions, numbers, and sets as well as grains of sand; i.e. that it is equivalent in meaning to (rather than merely, has as its abstract counterpart) the claim 'there is a 1-1 function from the set of natural numbers to the set of grains of sand'. Surely this is implausible, for surely it is consistent to maintain that there are infinitely many grains of sand but no numbers of functions or sets.[54]

I have argued that use of the quantifier 'there are only finitely many' or its negation does not necessarily involve commitment to abstract entities; and the same could of course be said for the 'fewer than' quantifier. Analogously, I argued in Chapter 4 that to use the complete logic of Goodmanian sums in one's theories does not necessarily involve commitment to abstract entities. In all these cases however, the question remains whether it might not be better to replace the theory that invokes the extra logic by another theory that invokes abstract entities but does without the extra logic.

That is exactly what the *first-order platonist* advocates doing. By a first-order platonist I mean someone who accepts theories that postulate abstract entities, but doesn't accept any logic beyond first-order logic. Such a first-order platonist has no resources at his disposal with the power of cardinality quantifiers like 'there are only finitely many F's' or 'there are fewer F's than G's'. For instance, whatever non-logical vocabulary he introduces and whatever consistent and recursively axiomatized (or recursively enumerably axiomatized) set T of axioms he asserts involving this vocabulary, there will be truths involving the cardinality quantifier that (when translated into the language of T) do not follow from T; and even if we were to give up the restriction that T must be recursively axiomatized (or recursively enumerably axiomatized), there will be valid

---

[54] I have shifted from 'there are only finitely many grains of sand' to 'there are infinitely many grains of sand' simply because the sense in which the abstract counterpart of the former is ontologically committed to functions etc. is less obvious than the sense in which the abstract counterpart of the latter is (since the abstract counterpart of the former is a *denial* of an existence-claim about functions). (There is however still a clear sense in which the abstract counterpart of the former commits one to functions etc.: only in the context of a theory that asserted the existence of lots of functions could the claim about the non-existence of a 1-1 function from the set of natural numbers to the set of grains of sand serve as an abstract counterpart of the claim that there are infinitely many grains of sand.) When I use the quantifier 'there are only finitely many' in the comparison of products, it is to assert finitude, not to assert infinitude.

inferences involving the cardinality quantifier that (when translated into the language of T) are not validated by T.[55] But this does not disturb the first-order platonist: the first-order platonist rests content with a recursively axiomatized theory—say, first-order set theory—in which we can translate the cardinality quantifier in such a way that *an important part of* its content is captured. (The same translation into *second-order* set theory would give the *full* content of the cardinality quantifier; but of course second-order set theory has a non-compact and non-recursively enumerable logic, so using a translation of the cardinality quantifier into this theory would be no gain.)[56] *We know by experience* that the platonistic first-order weakening of the cardinality quantifier suffices for physics. (It suffices for classical mathematics, which in turn suffices for physics.) So the first-order platonist has a method for *doing without* cardinality quantifiers in physics and *replacing them by* weaker set-theoretic surrogates in a compact and recursively enumerable logic.

This way of looking at things reinforces the point I made earlier, that use of the cardinality quantifier isn't platonistic, what's platonistic is only a certain set-theoretic *surrogate* for the cardinality quantifier. But it also shows a *prima facie* advantage of platonism: if we *do* use the first-order set-theoretic surrogate for the cardinality quantifier (and a similar first-order set-theoretic surrogate for the logic of Goodmanian sums), then we can make do in our theorizing with a compact and recursively enumerable fragment of logic. And isn't that an advantage?

I must admit that I think that it *is* an advantage. Consequently, I think it would be highly desirable to show that the nominalist too can do without the cardinality quantifier and the complete logic of Goodmanian sums, and can make do instead with weaker surrogates in a compact and recursively enumerable logic. I'm inclined to think in fact that this claim is true, and will give some considerations in support of this shortly; but I don't believe that the claim can be proved without a great deal of work

---

[55] I say that an inference is *validated* by a theory T if the conclusion of the inference follows from the premises of the inference together with the premises of T.

[56] The reason that the translation into first-order set theory doesn't give the full content of the cardinality quantifier is that there are models of first-order set theory that aren't models of second-order set theory—viz. the non-standard models of first-order set theory—and in many of these non-standard models an infinite set can satisfy the set-theoretic formula which 'says that' it is finite.

(which I haven't done), so the question arises, what if it *isn't* true? Would nominalism thereby be defeated?

To this I think the answer is no. For although there are certainly advantages to using only a compact and recursively axiomatized fragment of logic in developing physics, there are also advantages to keeping one's ontological commitments to a minimum; and the situation that we would be in (on the assumption that nominalism can't be made to work without going beyond first-order logic) is that we would have to make a choice as to which of two desirable goals is more important. It seems to me that the methodology to employ in making such decisions is a holist one: we should be guided by considerations of simplicity and attractiveness of overall theory. It seems totally unreasonable to insist on sticking to the requirement that logic be kept compact and recursively enumerable, *whatever* the costs for ontology; it is the simplicitly of the *overall conceptual scheme* that ought to count (as Einstein pointed out long ago against those who thought that the simplicity of Euclidean geometry should lead us to stick to it come what may).[57]

Admittedly this does not settle the issue of whether one should be a nominalist in the case at hand—the case where you can maintain nominalism by using a cardinality quantifier together with the complete logic of Goodmanian sums but (we are supposing) can't maintain nominalism with a weaker logic. All I've said is that one must look to overall theory to decide. My own view is that even if we are ultimately forced to make this decision—that is, even if it turns out that there is no reasonable way for the nominalist to make do with first-order logic—then nominalism is the reasonable position. For in the first place, the broader logics under consideration have their attractive aspects as well as their drawbacks.[58]

---

[57] The analogy here is not perfect: after all, Einstein was proposing a revision of geometry in the sense that some of the formerly held geometric claims were to be given up; whereas in the present case we are not considering a revision of logic in this sense, but merely an expansion of what counts as logic. Since, however, expansions are less radical than revisions, it is all the more inappropriate to resist expansions if such expansions will simplify one's total theory.

Note that I do not claim that logic is always to be expanded to keep down ontology. If the only way to nominalize the Newtonian theory of gravitation were to introduce a 'quantifier' Q such that $QxF(x)$ meant that there is something which is F and which is part of a universe that obeys the laws of Newtonian gravitational theory, then I would certainly conclude that Quine's argument for platonism was successful.

[58] See for instance Montague 1965, and the last two paragraphs of this chapter.

And in the second place (and more importantly) the use of this fairly small amount of extra logic saves us from having to believe in a large realm of otherwise gratuitous entities, entities which are very unlike the other entities we believe in (due for instance to their causal isolation from us and from everything we experience), and which give rise to substantial philosophical perplexities because of these differences.[59] In this situation, to insist on sticking to first-order logic because it is compact and recursively enumerable seems to me a bit like insisting on sticking to monadic logic because it is decidable.[60]

II

I will now say something about the question of how good the prospects are for making do with first-order logic. We have seen in earlier chapters how to give a nominalistic theory N in a broader logic that is an adequate nominalistic formulation of the Newtonian theory of gravitation; it seems reasonable then, in looking for a first-order nominalistic theory of gravitation, to look at first-order subtheories of N.

It should be noted at the outset that there is a quite uninteresting way to get a first-order subtheory of N with precisely the same first-order sentences (i.e. sentences not containing the cardinality quantifier or any second-order quantifiers) as consequences that N has as consequences: simply take as axioms all the first-order sentences that follow from N. This, however, is a bit reminiscent of the idea of dispensing with electrons in an axiomatization of physics by taking as axioms all those consequences of our standard theory that don't contain references to electrons. Let us then reject this strategy, on the grounds that it does not yield a sufficiently *attractive* first-order theory.

Since 'attractiveness' is not an easy notion to formalize, it seems that the only workable strategy in investigating whether there is an attractive first-order subtheory of N that is adequate to physics is to look at some particular first-order subtheory of N that does seem attractive, and try to

---

[59] See for instance Benacerraf 1973, Hart 1977 (especially pp. 123–7), and Jubien 1977. Also, for problems of quite a different sort, Lear 1977, Putnam 1977, and Benacerraf 1965.

[60] This last piece of hyperbole was suggested to me by remarks in Boolos 1975 and Tharp 1975. Both these papers, along with Tarski 1936, raise important issues about the decision of what to count as logic.

prove that it is sufficiently powerful. Of course, if it turns out that the particular subtheory one has investigated is *not* powerful enough, that won't prove that *no* attractive first-order subtheory of N is powerful enough: one may have simply left out first-order axioms that should have been included. But there seems to be no way around this difficulty, without giving precise formal content to the notion of attractiveness; and doing *that* seems obviously impossible.

There is, however, a rather natural first-order subtheory of N to investigate, and I would conjecture that this subtheory (which I will call N_0) *is* sufficiently powerful, in the sense of being adequate for the development of standard gravitational theory. I believe that by investigating this conjecture we will be able either to substantiate it, or to find out enough about why it fails so that we will able to supplement the theory (in an attractive way) with additional first-order axioms so that the resulting theory will be sufficiently powerful.

To see what the first-order subtheory N_0 of N that I have in mind is like, let us first not worry about the 'fewer than' quantifier and worry only about how to eliminate the use of the second-order quantifier, i.e. of what I've called the complete logic of Goodmanian sums. Intuitively, these second-order quantifiers range over all regions that contain only points in the domain of the first-order quantifiers; the problem in finding an adequate first-order subtheory of N, then, is the problem of finding an adequate first-order nominalistic axiomatization of the notion of region.

Looking at the matter platonistically, the set of all regions forms a complete atomic near-Boolean algebra, where by a near-Boolean algebra I mean something just like a Boolean algebra except not containing a zero element. (Recall the earlier stipulation that there is to be no region that contains no space-time points.) The atoms of the algebra, i.e. the regions with no proper subregions, are of course just the space-time points: to say that the algebra is atomic is to say that every region contains such atoms (and this implies that every region is the sum of the atoms it contains). Now, there is no difficulty in giving a complete first-order axiomatization of the notion of an atomic near-Boolean algebra (using, say, the primitive '⊆', meaning 'is a subregion of'). The problem is with the notion of completeness; i.e. with the idea that there are as many regions as there possibly could be, given that there are only the space-time points that there are. The platonistic method of specifying completeness of the algebra is to say that for every non-empty set of space-time points, there is a region containing (i.e. having as subregions) the points in that set and no others. The obvious nominalistic tack is to replace this claim by an axiom schema: to regard as an axiom each sentence of form

(13) $\forall u_1 \ldots, u_n \{\exists x[x \text{ is a point and } \Theta(x, u_1, \ldots, u_n)] \rightarrow \exists r \forall x[x \text{ is a point} \rightarrow (x \subseteq r \leftrightarrow \Theta(x, u_1, \ldots, u_n))]\}.$

Here I am using only one style of variable, ranging over regions generally; 'x is a point' is defined as '$\forall y(y \subseteq x \rightarrow y = x)$'. $\Theta(x, u_1, \ldots, u_n)$ can be any formula in the nominalistic language: in particular, it can contain physical vocabulary like 'spatio-temporally between' or 'congruent in gravitational potential'; and it may include quantifiers that range over regions. (This latter stipulation means that some of the instances of our schema will make impredicative assertions of region-existence. This seems legitimate: on the realist approach to space-time I've adopted, regions are physical entities that objectively exist independently of our picking them out. But if you don't like impredicativity, you could weaken the theory by disallowing such instances of (13).)

Having axiomatized the notion of region in this first-order way, it is clear how to get a subtheory of N with no second-order quantifiers: simply take N and restrict all first-order quantifiers by the defined predicate 'point', then replace all second-order quantifiers by unrestricted first-order quantifiers, and append the axioms for regions.[61,62] [So in particular, the Dedekind continuity claims (for the geometric ordering of space-time points, and for the ordering of space-time points with respect to each scalar) are each made by a single axiom rather than by a schema: a common notion of region is used in axiomatizing each of the Dedekind continuity claims. This turns out to be important for insuring that the different orderings interrelate in the desired way.] This theory (I'll call it N*) isn't quite a first-order theory, because it still contains the cardinality quantifier. So we now have to get rid of that.

---

[61] Here for simplicity I'm assuming that N has been written so that particles are not explicitly quantified over in formulating it—or if you like, particles are identified with their trajectories. (Cf. note 51.) Obviously this is not essential to the strategy in the text of how second-order quantifiers are to be eliminated, it just makes that strategy a bit easier to describe.

[62] This is still a subtheory of N despite its additional vocabulary '$\subseteq$', because the new vocabulary was definable in N using second-order quantification. The same will go for the predicate '<' to be introduced later.

LOGIC AND ONTOLOGY 101

When I sketched how to formulate N in Chapter 8, I made use of the 'fewer than' quantifier $\mathcal{F}_0$; but as I remarked, this can be dropped in favor of the finiteness quantifier $\exists_{fin}$, if we add a predicate to the theory. It is necessary to show how this is done, before going on to perform the further task of dropping '$\exists_{fin}$'.

The predicate we need to add, in order to replace $\mathcal{F}_0$ by $\exists_{fin}$, is a binary predicate '$\leq$' holding between regions: '$r_1 \leq r_2$' is to mean intuitively that r_2 contains no fewer points than does r_1 (with the convention on 'fewer than' introduced before: all infinite regions are to be regarded as containing equally many points). Since the 'fewer than' quantifier was applied in Chapter 8 only in the context of points rather than of arbitrary regions—that is, it occurred only in context $A(x)\mathcal{F}_0 B(x)$ in which the formulas $A(x)$ and $B(x)$ couldn't be satisfied by anything other than points, in any model of the theory—then it is clear that '$\mathcal{F}_0$' can be dropped in favor of '$\leq$' as long as we can axiomatize '$\leq$' in such a way that in any model, '$x_1 \leq x_2$' will be satisfied by precisely the pairs $\langle r_1, r_2 \rangle$ such that r_2 contains no fewer points than r_1.[63] It is in order to meet this condition that the quantifier $\exists_{fin}$ must be introduced.

An axiom system that meets this condition is as follows. (The theory of atomic near-Boolean algebras is presupposed as a background theory. Although the only primitive I've introduced for that theory is '$\subseteq$', I'll use the defined term 'point' introduced above and various other Boolean notions like '$\cup$', since it is obvious how to paraphrase claims involving them in terms of '$\subseteq$'. The only additional primitive to be used in the axiom system is the cardinality relation '$\leq$'; '$<$' and '$\approx$' are defined in terms of it, i.e. '$x < y$' is defined as '$x \leq y$ and not $y \leq x$' and '$x \approx y$' as '$x \leq y$ and $y \leq x$'. Also, 'Inf(x)', meaning intuitively that x has maximum size, is defined as '$\forall y (y \leq x)$'; Axiom 5, together with the others in 1–7, ensures that this amounts to x being infinite.) The system consists of Axioms 1–7 below:[64]

---

[63] In fact, since in Chapter 8 we applied the cardinality quantifier only to points in equally spaced regions, we could be satisfied with axioms that guaranteed that the claim in the text held for any equally spaced regions r_1 and r_2. This fact is of relevance in connection with note 66.

[64] *Note added to 2nd edition:* A more pellucid axiomatization would have included the axiom

    0. $\qquad\qquad\qquad x \subseteq y \to x \leq y$.

Then 7_1 (introduced in the next paragraph) would have been redundant given the mereological background even without Axiom 7, since Axiom 0 would guarantee that the region V containing all space-time points satisfies the definition of Inf.

1. $x \leq y \wedge y \leq z \rightarrow x \leq z$
2. $x \leq y \vee y \leq x$
3. $\exists x \exists y (x \not\leq y)$
4. $\text{Point}(x) \rightarrow \forall y (x \leq y)$
5. $\text{Inf}(x \cup y) \rightarrow \text{Inf}(x) \vee \text{Inf}(y)$
6. $\neg \text{Inf}(x) \wedge \text{Point}(y) \wedge y \not\leq x \rightarrow x < x \cup y \wedge \neg \exists z (x < z < x \cup y)$
7. $\neg \text{Inf}(x) \rightarrow \exists_{fin} y [\text{Point}(y) \wedge y \subseteq x]$.

What about eliminating $\exists_{fin}$? Axioms 1–6 don't contain it, and they together with[65]

6A. $\neg \text{Inf}(x) \wedge w \cup y \approx x \wedge \text{Point}(y) \wedge y \not\subseteq w \rightarrow w < x \wedge \neg \exists z (w < z < x)$

7_1. $\exists x\, \text{Inf}(x)$

(which also don't contain '$\exists_{fin}$' but only the defined predicate 'Inf') are enough to guarantee (in the context of the axioms for an atomic near-Boolean algebra) that in any model of the theory, if Eq is the extension of the formula '$x_1 \leq x_2 \wedge x_2 \leq x_1$' in the model then:

(a) Eq is an equivalence relation whose equivalence classes form a linear ordering with first and last elements.
(b) Each equivalence class other than the last has an immediate successor.
(c) The last equivalence class has predecessors but no immediate predecessor; but it is the only equivalence class with this property.
(d) For each positive integer n, the nth equivalence class in the ordering contains precisely those regions that contain exactly n points.

So the last equivalence class in the ordering—the one that contains precisely the regions satisfying the predicate 'Inf'—contains only infinite

---

Axiom 0 (and hence 7_1) does follow from 1–7. For if Inf(y) then $x \leq y$ by definition; so we need only deal with the case where ¬Inf (y), and hence where, by 7, there are only finitely many points in y that aren't in its subset x. The proof is by induction on the number n of such points. If n is 0, x is just y, and $y \leq y$ is guaranteed by 2. For the induction step, suppose that whenever z is a subset of y with k fewer members than y, $z \leq y$, and that x is a subset of y with $k+1$ fewer members. Letting p be one of those members, the induction hypothesis gives that $x \cup p \leq y$, and part of 6 gives that $x \leq x \cup p$, so 1 gives $x \leq y$ as desired.

[65] *Note added to this edition*: The original edition erroneously omitted 6A. I have had to do a bit of rewriting of the remainder of this section to correct this, but have stuck to the original edition as closely as possible. 6A follows from 1–7 by an inductive argument similar to that used for Axiom 0 in the previous note.

regions, and there are no infinite regions in the first, second, third, etc. equivalence classes. However, a compactness argument shows that there are models of axioms 1–6 and 6A and 7_1 in which there are 'non-standard' equivalence classes, equivalence classes which occupy no finite position but also are not last; regions in these equivalence classes will be infinite, but will not satisfy the defined predicate 'Inf'.[66]

If we are to rule out such non-standard models of '≤', we must replace axioms 6A and 7_1 by axiom 7 (which in conjunction with 1–6 entails them). That this strengthening does rule out non-standard equivalence classes is clear: it implies that the ordering of equivalence classes is a well-ordering, and this together with (a)–(c) implies that the ordering has order type $\omega + 1$ (i.e. of the positive integers followed by one infinite element). This last result plus (d) shows that in each model of the axioms of atomic near-Boolean algebra plus axioms 1–7, '≤' is satisfied by precisely the pairs $\langle r_1, r_2 \rangle$ such that r_2 has no fewer points than r_1.

We could also rule out the non-standard models by a second-order induction axiom: either

7′. $\forall P[\exists x Px \wedge \forall x \forall y (\text{Point}(y) \wedge Px \to P(x \cup y)) \to \forall x (Px \vee \text{Inf}(x))]$,

or if we also include Axiom 6A, the somewhat simpler

7′*. $\forall P[\exists x Px \to \exists x(Px \wedge \forall w(w < x \to \neg Pw))]$.

These second-order axioms are not nominalistic: they involve second-order quantifications that go beyond the complete logic of Goodmanian sums, for the predicate-quantifier ranges over predicates of *regions that aren't points*. But they are of interest, because they have obvious first-order weakenings 7″ and 7‴*: simply replace them by first-order induction schemas. For instance, 7′* is

---

[66] Note that no such equivalence classes could contain any equally spaced regions if space-time was Archimedean. The Archimedeanness of space-time was a consequence of the original axiomatization of its geometry, the axiomatization using the complete logic of Goodmanian sums; this fact together with note 63 is sufficient to show that the cardinality quantifiers aren't needed when one has the complete logic of Goodmanian sums. But now that we've dropped the use of the logic of Goodmanian sums, there will be non-standard models of space-time in which it has non-Archimedean structure, so there is no guarantee that the non-standard equivalence classes can't contain equally spaced regions in some models.

7‴*. $\forall u_1, \ldots, u_n [\exists x \Phi(x, u_1, \ldots, u_n) \rightarrow \exists x \{\Phi(x, u_1, \ldots, u_n) \wedge \forall w(w < x \rightarrow \neg \Phi(w, u_1, \ldots, u_n))\}]$.

(7″ entails 6A and 7‴* doesn't; but given 6A, they are equivalent.) The models of the theory consisting of axioms 1–6 plus 7″ (or 1–6 plus 6A plus 7‴*) have no 'discernible' non-standard equivalence classes; and the theory seems to be a very natural first-order weakening of axioms 1–7.

If we combine 1–6 and 7″ with the rest of our theory N*, we obtain a completely first-order subtheory N_0 of N. In doing this, we are to let the axiom schema 7″ take as instances any formula in the language of N_0—that is, it can contain empirical vocabulary like 'Temp-Bet' as well as '$\leq$' and '$\subseteq$'. Also, we are to expand the instances of schema (13) to include formulas containing '$\leq$'.

III

N_0 is a rather natural first-order subtheory of N to look at, in that it results from something equivalent to N simply by replacing two second-order statements (viz. axiom 7 and the second-order strengthening of (13)) by schemas. This system, then, is related to the second-order theory N in very much the way that first-order set theory is related to second-order set theory: there too, we get the first-order weakening of the theory simply by taking a second-order axiom (in this case the replacement axiom) and making a first-order schema out of it. This analogy might lead us to suspect that just as N has all the nominalistic consequences that platonistic formulations of Newtonian gravitation theory have in the context of second-order set theory, so too N_0 will have all the nominalistic consequences that platonistic formulations of Newtonian gravitation theory have in the context of first-order set theory.

It would be nice if this guess were correct, but I don't think that it can be. For something analogous to first-order number theory appears to be imbeddable in the system N_0 (using the points in an arbitrary infinite equally spaced region with one endpoint, instead of the natural numbers). Consequently, N_0 ought to have a Gödel sentence expressible but not provable in it; and this Gödel sentence ought to be provable in the system P_0 of first-order platonistic gravitational theory.[67] If this argument-sketch

---

[67] Essentially this observation was made to me by John Burgess and Yiannis Moschovakis.

is correct then there will be some very *recherché* consequences of P_0 that are expressible but not provable in N_0. Still, I suspect that the extra strength that P_0 has over N_0 is confined to such *recherché* consequences; N_0 is I suspect sufficient for all nominalistic consequences we would normally be interested in deducing from P_0, and more than sufficient for developing the usual theory of gravitation. Compare Peano arithmetic, the first-order theory that results from full second-order arithmetic by replacing the second-order induction axiom by a first-order schema: although second-order arithmetic has all the arithmetic consequences that arithmetic in the context of second-order set theory has, Peano arithmetic is weaker than arithmetic in the context of first-order set theory. Still, Peano arithmetic is strong enough for any ordinary arithmetical consequences. I would suspect that something analogous is true for N_0: that it is more than strong enough for any ordinary developments in the usual theory of gravitation. However, this is not a matter I have investigated very far, and I will leave it to others more adept at these matters than I to confirm or refute my suspicion. (As I've mentioned, if the suspicion turns out to be false, I would look for a natural first-order strengthening of N_0.)

A platonist might argue that even if I am right about the strength of N_0, nominalism is still in trouble: for since N_0 is weaker in nominalistic consequences than the first-order platonistic theory P_0, then it doesn't have all the nominalistic consequences that we ought to want. That is, we ought to want *all* the nominalistic consequences of P_0, even the very *recherché* ones that no one is interested in in practice, like the Gödel sentence of N_0. There is a certain plausibility to this argument for the inadequacy of N_0; but it doesn't seem to me that it can be used to support the platonistic theory P_0, it can only be used to support a second-order theory like N. For P_0 too (assuming that it is formulated in a recursively axiomatized system like Zermelo-Frankel set theory, or at least a recursively enumerably axiomatized system) has a Gödel sentence which is intuitively true, and by adding that sentence to P_0 we will get *recherché* consequences not obtainable from P_0 alone; these *recherché* consequences seem just as intuitively desirable as the Gödel sentence of N_0. And the same point holds not only for P_0 but for any expansion of P_0 with a recursively enumerable set of axioms. The point is that a price of restricting oneself to first-order logic (and to recursively or recursively enumerably axiomatized theories; but there seems to be little point in a restriction to first-order logic if one is going to allow the use of theories with no recursively enumerable axiomatization) is that one has to settle for a rather arbitrarily restricted theory. That is, for any first-order theory one settles for, there is a better one, one that seems intuitively to be true if the original one is, and is more powerful. This holds whether the first-order theory one settles for is a nominalistic one like N_0 or a platonistic one like P_0; hence it can't be used as an argument for the inadequacy of N_0 unless platonistic first-order theories are also admitted to be inadequate. Consequently, if one is committed to first-order theories, then the only obvious way to decide if one is good enough is to decide whether it is powerful enough to get the results that are seriously needed in practice, i.e. excluding *recherché* results like those obtained by Gödelization. As I've said, I think it highly likely that N_0 or some slightly stronger first-order subtheory of N passes this test.

The argument at the beginning of the previous paragraph, then, may indicate an inadequacy in N_0; but if so, it is an inadequacy in P_0 as well, and hence it is not an argument for platonism. If you want to cure this 'inadequacy', the only recourse is to go to a second-order theory—either N, or platonistic gravitational theory in the context of second-order set theory. But since as we've seen N has all the nominalistic consequences that second-order platonistic set theory has, it is hard to see in the context of second-order logic what the advantages of platonism can be. Either way, then, it looks as if nominalism triumphs.

Bibliography for Original Text

[1] Benacerraf, Paul 1965. "What Numbers Could Not Be". *Philosophical Review* 74: 47–73.
[2] Benacerraf, Paul 1973. "Mathematical Truth". *Journal of Philosophy* 70: 661–79.
[3] Boolos, George 1975. "On Second Order Logic". *Journal of Philosophy* 72: 509–27.
[4] Chihara, Charles 1973. *Ontology and the Vicious Circle Principle*. Ithaca, NY: Cornell University Press.
[5] Earman, John 1970. "Who's Afraid of Absolute Space?" *Australasian Journal of Philosophy* 48: 287–319.
[6] Friedman, Michael 1981. *Foundations of Space-Time Theories*. Princeton, NJ: Princeton University Press.
[7] Goodman, Nelson 1972. *Problems and Projects*. Indianapolis, IN: Bobbs-Merrill.
[8] Hart, W. D. 1977. "Review of Mark Steiner *Mathematical Knowledge*". *Journal of Philosophy* 74: 118–29.
[9] Hawking, S. and Ellis, G. 1973. *The Large-Scale Structure of Space-Time*. Cambridge: Cambridge University Press.
[10] Hilbert, David 1902. *Foundations of Geometry*. LaSalle, IL: Open Court, 1971.
[11] Jech, Thomas 1973. *The Axiom of Choice*. Amsterdam: North-Holland.
[12] Jubien, Michael 1977. "Ontology and Mathematical Truth". *Nous* 11: 133–50.
[13] Krantz, D., Luce, R. D., Suppes, P., and Tversky, A. 1971. *Foundations of Measurement*, vol. 1. New York: Academic Press.
[14] Lear, Jonathan 1977. "Sets and Semantics". *Journal of Philosophy* 74: 86–102.
[15] Montague, Richard 1965. "Set Theory and Higher Order Logic". In J. N. Crossley and Michael Dummett (eds), *Formal Systems and Recursive Functions*. Amsterdam: North-Holland, pp. 131–48.
[16] Putnam, Hilary 1967. "The Thesis that Mathematics is Logic". In his *Philosophical Papers*, vol. 1. Cambridge: Cambridge University Press, 1975, pp. 12–42.
[17] Putnam, Hilary 1970. "On Properties". In his *Philosophical Papers*, vol. 1, pp. 305–22.
[18] Putnam, Hilary 1971. *The Philosophy of Logic*. New York: Harper.
[19] Putnam, Hilary 1975. "What is Mathematical Truth". In his *Philosophical Papers*, vol. 1, pp. 60–78.

[20] Putnam, Hilary 1977. "Models and Reality", Presidential Address to the Association of Symbolic Logic. (Since published in his *Philosophical Papers*, vol. 3. Cambridge: Cambridge University Press, 1983, pp. 1–25.)
[21] Szczerba, L. W. and Tarski, A. 1965. "Metamathematical Properties of Some Affine Geometries". In Y. Bar-Hillel (ed.), *Logic, Methodology and Philosophy of Science 1964 Conference*. Amsterdam: North-Holland, pp. 166–78.
[22] Tarski, Alfred 1936. "On the Concept of Logical Consequence". In his *Logic, Semantics and Metamathematics*. Oxford: Clarendon Press, 1956, pp. 409–20.
[23] Tarski, Alfred 1959. "What is Elementary Geometry?" In L. Henkin, P. Suppes, and A. Tarski, *The Axiomatic Method*. Amsterdam: North-Holland, pp. 16–29.
[24] Tharp, Leslie 1975. "Which Logic is the Right Logic?" *Synthese* 31: 1–21.

Index

absolute rest 36, 47–9, 51
absolute space, *see* space-time (realist attitude toward); *see also* absolute rest
abstract counterpart P5; 22, 24, 26, 28–9, 94–5
affine geometry and affine transformations P5–P6, P7, P44–P45, P49; 49, 51–4, 63–5, 67, 71, 75n., 76, 90
a prioricity of mathematics 12–13, 15–16, 32–3
arbitrary choices P7; v, 32, 33, 44–9, 51n., 54, 63, 68–9, 79, 81
arithmetic P2–P3, P8–P12, P15; iv, 22–5, 105
Arntzenius, Frank P7, P41, P45

Baker, Alan P31, P37
Baker, David P41–P44, P46
Balaguer, Mark P2, P38n.
Benacerraf, Paul 98n.
Boolos, George 98n.
Burgess, John P4, P7, P15, P24, P44–P46; iii, 104n.

cardinality quantifiers P2, P8–P13, P15
  see also fewer; finiteness
Carnap, Rudolph P3
causal relevance P33–P34, P36; 36n., 43–4
chance P7
Chihara, Charles 5n., 45
Church's Thesis 14n.–15n.
coding (avoiding) P7
Colyvan, Mark P31, P37
comparativism about quantities P41–P44; 55–6
complete logic of the part/whole relation (= complete logic of Goodmanian sums) P12–P14, P23–P24, P29–P30; 38–9, 92–3
  dispensing with P13–P15, P22–P23; 99–101, 104–6

completeness (of a Boolean algebra) P22; 99–100
configuration space and phase space P6–P7, P33, P37
'confirmational holism' P31–P32
conservativeness P8, P12, P15–P30, P32, P35, P38; iv, 10–21, 22, 25, 30, 40, 41n., 44, 90
  ω-conservativeness P24–P25, P30
consistency of mathematics P18, P25; iv, 13, 14n., 15, 17, 18–21, 25, 30, 40, 41n.
conventional choices, *see* arbitrary choices
convention, truth by P38–P39
Craig's Theorem P19n; 8, 42, 43, 47

Dasgupta, Shamik P42
Dedekind continuity P14–P15, P27–P28; 32, 38, 58n., 100
Dorr, Cian P6n., P7, P35n., P41, P45, P47
Dummett, Michael P19–P20
Dürr, Detlef P6n.

Earman, John 36, 48
Einstein, Albert 97
electromagnetism 35, 36n., 43
elementary theories in the sense of Tarski P45; 40
Ellis, George 76n.
'error theories' P3–P4

fewer P9, P11, P23, P44–P45; 66, 68n., 92–5
  dispensing with notion of P44–P45, 93, 101, 103n.
fields P5, P40, P41, P49; 2, 35–6, 54, 55
finiteness P11–P13, P23, P28–P29; 93–5, 96n., 101–3
  dispensing with notion of P11, P23, P44–P45; 93, 104–6
finitism 3–4, 31–2, 37
Friedman, Michael 36, 48, 51n.

generalized Galilean
  transformation 50–4, 59, 71
general relativity, *see* relativity, general
  theory
Gödel sentences P28; 104–6
  see also conservativeness,
    ω-conservativeness
Goodmanian sums P12n.–P13n., P22, 37
  see also complete logic of the
    part/whole relation
gravitational constant, value
  of P41–P44, P46; 79
gravitational field intensity vs.
  gravitational potential as
  primitive 75n.–76n.

Hart, W.D. 98n.
Hawking, Stephen 76n.
Hellman, Geoffrey P2, P30n., P47
Hilbert, David P14, P22, P29; v, 27–9, 31,
  33, 35–40, 42–3, 46, 49–51, 56
Hilbert space P6n.
Hodes, Harold P9–P11
homomorphism 26, 28, 89

idealization in science P5, P33, P43; 81n.
impredicativity P9–P11, P13n., P14, P22,
  P24, P45; i–ii, 45, 100
indispensability arguments P1, P5,
  P30–P38; iii–iv, 2, 5, 8
  for non-fundamental science P5, P6,
    P33–P34
informal proofs 14n.–15n.
integration P15, P40–P41; 2
intrinsic facts and intrinsic
  explanations P4–P7, P10n.,
  P14–P16, P33, P36, P40; ii–iii, v, 29,
  44–7, 49–50, 51n., 55–7, 63
invariance P6n., P41; 29, 33, 50–3,
  56, 59–60, 64–5, 67, 69–71, 76,
  78–81, 86

Jech, Thomas 19n.
joint axiom system (JAS) 59, 61
Jubien, Michael 98n.

knowledge of mathematics i–ii, 4,
  32, 98n.
Krantz, David 57–9, 79n.
Kripke, Saul P10n., P25n., P44
Kronecker, Leopold 25

Lear, Jonathan 98n.
Leng, Mary P31, P37
logic P2, P9–P12, P23–P30, P32n.,
  P39, P45; 16–17, 23–4, 38–9, 92–8,
  105–6
  prospects for nominalist making do
    with first order P11–P15, P22–P23;
    vi, 98–106
Luce, R. Duncan 57–9, 79n.

Mach, Ernst 48
Maddy, Penelope P31, P37
Malament, David P5–P6
Malink, Marko P21n.
mathematical intuition 5n.
mathematics, pure vs. impure P18–P21,
  P26, P47; 8–10, 11n.–12n., 13, 18,
  20–1, 25n.
'mathematics' and 'mathematical theory',
  my use of the terms P17–P18,
  P20n.; iin., 2n.
measurement theory P7n., P36; 57–9
Melia, Joseph P31, P37
Mereology, *see* complete logic of the
  part/whole relation; elementary
  theories
metalogic P1, P5, P12, P16n., P23–P24,
  P33; 5, 6n.
Meyer, Glen P6n.
mixed comparative primitives
  P44–P46
modality as surrogate for mathematical
  entities P46–P48
Montague, Richard 39n., 97n.
Moschovakis, Yiannis 104n.
Mundy, Brent P7, P15, P42n., P46

necessary truth of mathematics P16,
  P18–P19, P25; 12–13, 15–16
Newton, Isaac 48
nominalism P2, P7; iii, 1–5, 31–40, 53–6,
  90–1, 92–8, 105–6
  nominalistic assertions P17; 10–11
  'revolutionary' vs.
    'hermeneutic' P3–P4
non-arbitrariness of mathematical
  axioms P2, P3, P9; 4–5, 15–16, 25
non-measurable regions P27–P28; 37n.
non-necessary axioms 58n.
non-standard models P14–P15, P44;
  39–40, 96n., 103–4

objectivity of mathematics P2–P3, P30, P37–P38; 15–16
ontological commitment P1–P3, P10, P31–P33, P35; ii–iii, vi, 37–9, 93–8
operationalism 3–4, 31–2, 37, 57

Perry, Zee P46
platonism involving more than mathematical existence? P2, P4, P30, P37–P39
platonistic arguments, use of in establishing possibility of nominalism 5, 6n., 14n.
point particles P41; 81n., 85, 100n.
  point masses P45–P46
Principles C, C′, and C″ P21n.; 11–12
properties P7, P9n., P10–P11, P12n., P46; 55
Putnam, Hilary P2, P30, P31; iii, 5n., 25n., 55, 98n.

quantum mechanics P6–P7
quasi-inductive knowledge of logical results 13, 14n.–15n.
Quine, Willard V. O. P1, P3, P7, P30–P32, P39, P54–P56; 2, 5, 97n.

Rayo, Agustin P22
real-closed fields P14–P15, P29
real numbers, use of to represent spatial positions P14–P16, P38, P45; 3, 31–4
  see also representation theorems
regions other than points P10n., P12–P13, P22, P27–P28, P40, P44–P45, P48; 37–8, 62, 99–100
  see also elementary theories in the sense of Tarski
reinterpreting mathematics, vs. being a fictionalist or instrumentalist about mathematics P7n.; 1–2, 6, 25n., 45
relativity:
  special theory 47, 49, 63
  general theory P5–P6, P7, P48; 63, 75n.–76n.
representation theorems P4, P14–P15, P22–P23, P28–P29, P35–P37, P44, P45, P49; v, 26, 28–9, 40, 49–54, 56–9, 61–2, 64, 67–72, 75–6, 84, 89–90

Robinson's Theorem 18–19, 41
Rosen, Gideon P4, P15, P39n., P44

schemas, extensibility of P12–P13, P24, P26–P27
schemas vs. second-order axioms 39, 100, 103–5
second- (and higher-) order logic P9–P11, P13n., P15, P22–P24, P27–P29, P45; 38–41, 96–8, 103–6
  subsystems of second-order arithmetic P15
set theory:
  full vs. restricted 13, 17–19
  pure vs. impure P18–P21, P25–P26; 8–10, 13, 17–19
  second order 39–40, 96, 104–6
Shapiro, Stewart P24, P29, P37
Simpson, Stephen P15
Sober, Elliot P31, P32, P37
space-time, realist attitude toward P10n., P12–P13; 31–6, 58n.
Suppes, Patrick 57–9, 79n.
synthetic approach P15; 42–6
  see also arbitrary choices; intrinsic facts and explanations; invariance
Szczerba, L. W. 49, 51–2, 54, 63

Tarski, Alfred P10, P14, P28, P45; 40, 49, 51–2, 54, 63, 94, 98n.
tensor analysis P5, P36, P49; 48, 50–1, 75n.–76n., 84
Teufel, Stefan P6n.
Tharp, Leslie 98n.
theoretical indispensability, see indispensability arguments
truth of mathematics P3–P4, P25; i–ii, 4–5, 7, 14–16, 20, 22, 25, 30, 40, 98n.
  see also convention, truth by
Tversky, Amos 57–9, 79n.

uniqueness theorems v, 28–9, 46, 49–54, 56–9, 61
urelements P18; 9, 11, 17

Weinstein, Scott iii, 20, 41n.

Yablo, Stephen P22, P31, P35–P37